KENKEN

FOR EXPERTS ONLY

VOLUME 3

100 CHALLENGING LOGIC PUZZLES
THAT MAKE YOU SMARTER

CREATED BY
TETSUYA MIYAMOTO

KENKEN
www.kenkenpuzzle.com

To access free, unlimited puzzles of all sizes and difficulty levels, visit www.kenkenpuzzle.com

Download the free KenKen Classic app on any iOS, Android, or Kindle device in the iTunes App Store, Google Play Store, or Amazon App Store.

KenKen: For Experts Only, Vol. 3

www.kenkenpuzzle.com

Project editors: Ben Scaglione and Jerry March
Designer: Susannah Fears

ISBN-13: 978-1-945542-14-5
First Edition: January 2022

INTRODUCTION

WARNING:

The puzzles contained in this 3rd collection of expert puzzles have been "Kenerated," by our crack team of "Kenerators," to challenge the most expert of experts, the best of the best. If you conquer them, consider yourself a true KenKen aficionado, worthy of the title "Expert KenKen'er".

WE SALUTE YOU!

Developed in 2004 by renowned educator and puzzle master Tetsuya Miyamoto, KenKen has quickly become the world's most addictive and fastest growing puzzle since Sudoku. Miyamoto-sensei invented KenKen for his students in Yokohama, Japan, as a tool not only to improve mathematics skills, but also to encourage independent learning by emphasizing reasoning, creativity, and perseverance. He never anticipated his simple yet sophisticated puzzle would transcend his classroom and flourish into an international sensation. KenKen is now published daily in over 150 newspapers and magazines worldwide including The New York Times, The Times (UK), Spiegel Online (Germany), El Pais (Spain), The Globe and Mail (Canada), The Los Angeles Times, and many others. Over 50 million puzzles are played online each year, with hundreds of millions more solved in print and on mobile devices.

As an experienced KenKen solver, you have undoubtedly added to these totals. And while online play is great in the home or office and the newspaper accompanies your morning commute, KenKen books are perfect almost anywhere…planes and trains, bedrooms and bathrooms. We are pleased to present you with the book you now hold in your hands: KenKen: For Experts Only, Volume 3…the third in KenKen Puzzle Company's official series of KenKen puzzles for experts. We have taken great care to curate this special group of puzzles. They will captivate you, challenge you, and ultimately distinguish you as a KenKen master.

While it's true that KenKen was invented to educate and exercise the mind, its main goal is to entertain. So sit back, relax, grab a pencil and start solving. Enjoy!

How To Solve

At its core, KenKen is a simple yet rich logic puzzle with easy-to-understand rules:

◊ Fill each square in the grid with a single number. In a 9x9 grid, use the numbers 1 through 9.

◊ Do not repeat a number in any row or column. For example, in a 9x9 grid, each row and each column should be filled in with the numbers 1 – 9 with no duplication.

◊ Each heavily outlined set of squares is called a "Cage."

◊ The numbers in each Cage must combine – in any order – to produce the Target Number indicated in the top corner of the Cage by using the math operation next to the Target Number. For example, if the Target Number shows "18x" in a 3-square Cage, the numbers in that cage must total 18 in any order by using multiplication (9x2x1, 6x3x1, 3x2x3, etc).

◊ A number may be repeated within a Cage as long as it is not in the same row or column. You can do this when a Cage has an "L" shape or spans several rows or columns.

Hint: To start, fill in the single-square Cages first. Since no calculation is required and the answer is "given" to you, we call these "Freebies."

And that's it!

Solving a KenKen puzzle involves pure logic and mathematics. No guesswork will ever be needed, and each puzzle has only one solution. So sharpen your pencil, sharpen your brain, and find out why Will Shortz, NPR Puzzle Master and The New York Times Puzzle Editor, calls KenKen "The most addictive puzzle since Sudoku!"

Visit www.kenkenpuzzle.com any time to play unlimited puzzles of all sizes and difficulty levels, absolutely free of charge.

1

+

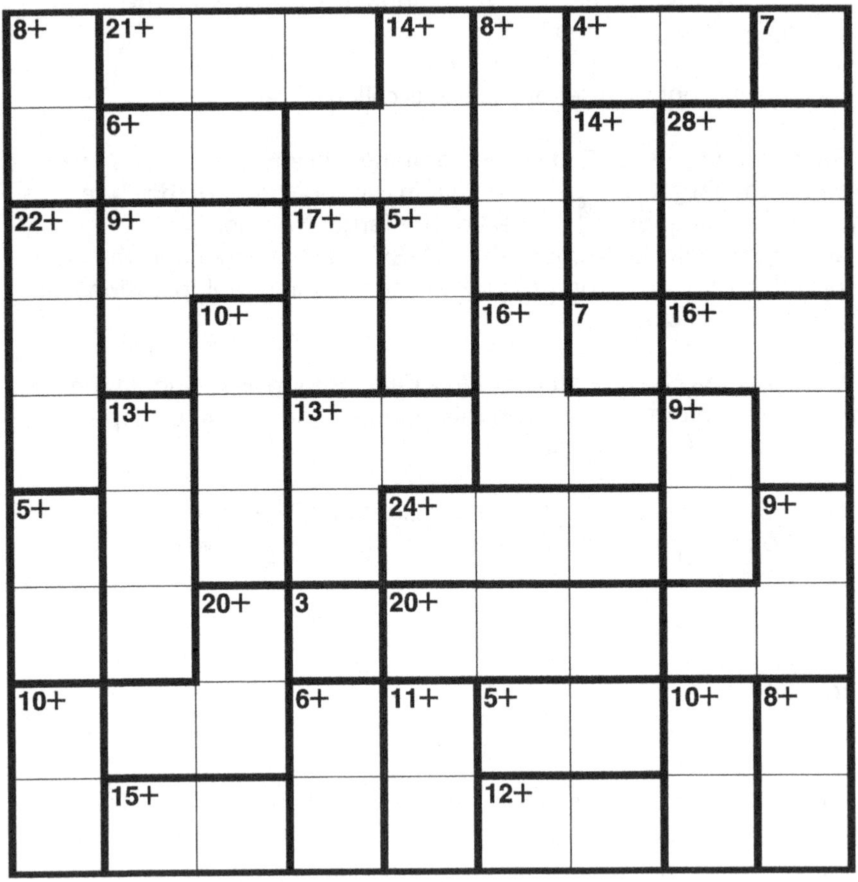

SOLVING
TIME

2

+

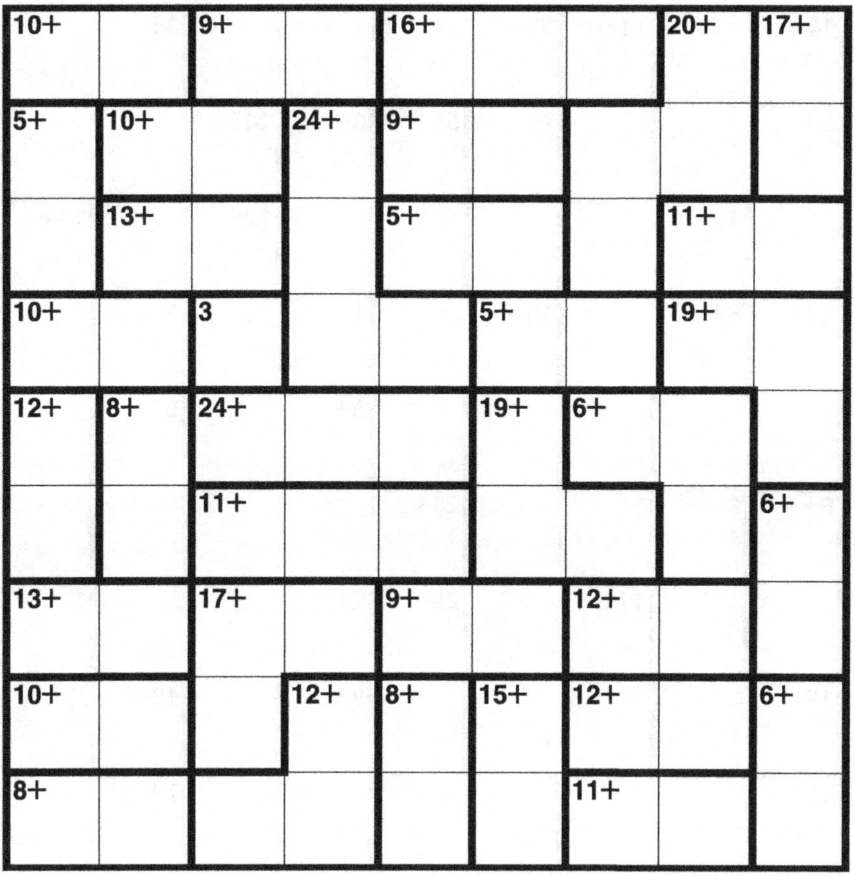

SOLVING
TIME

3

+

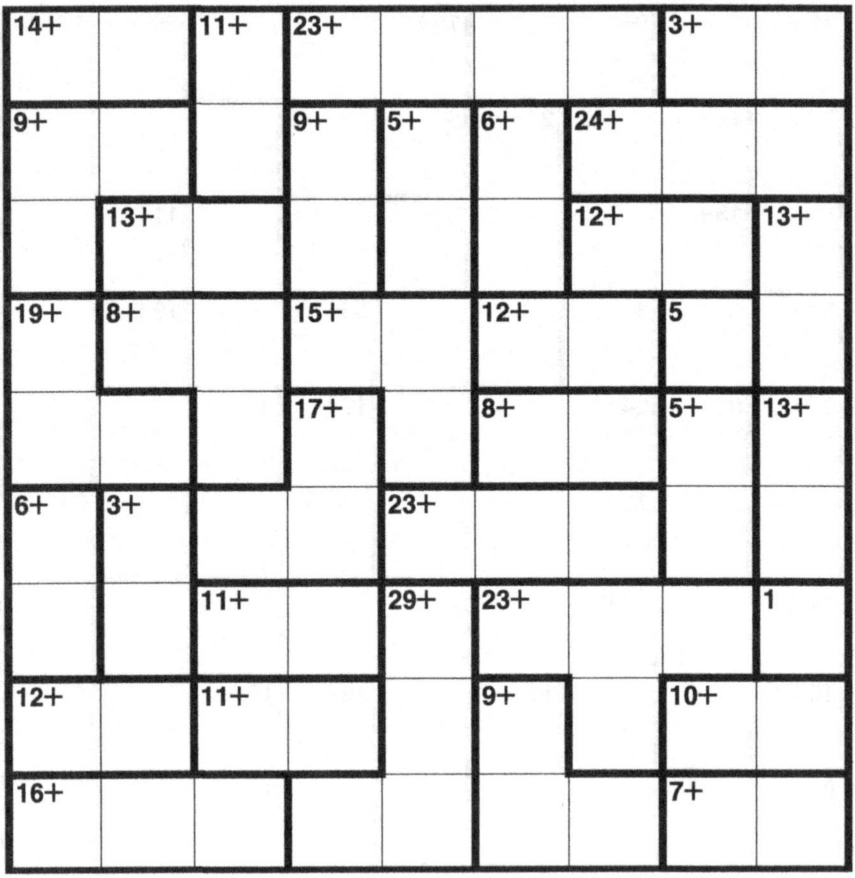

SOLVING
TIME

4 +

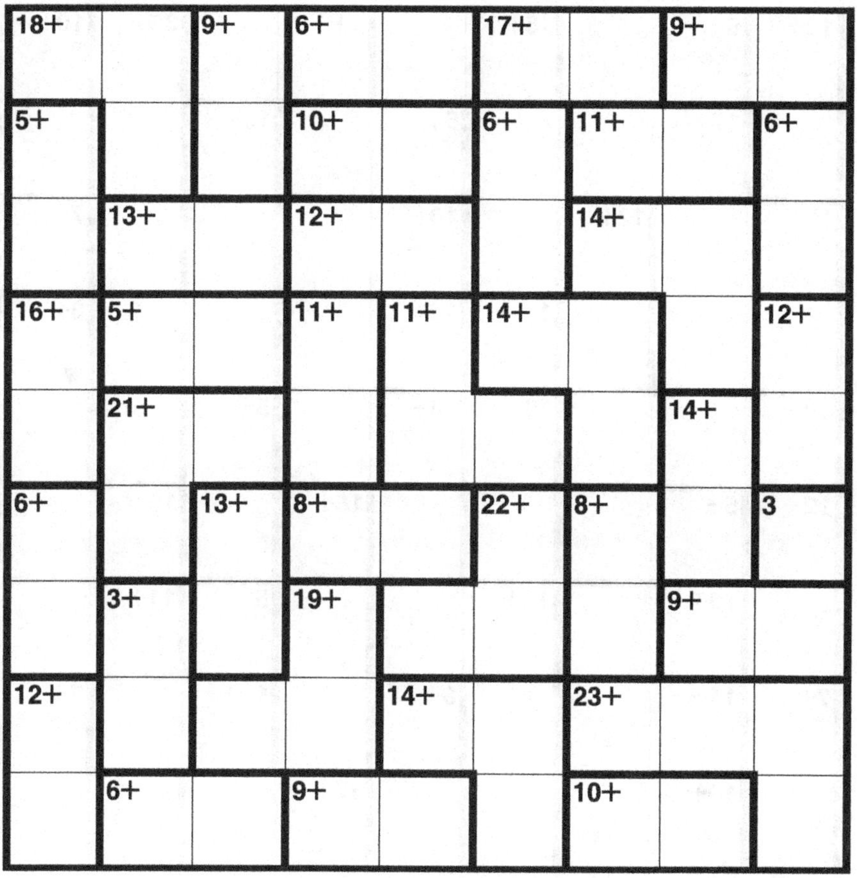

SOLVING
TIME

5

+

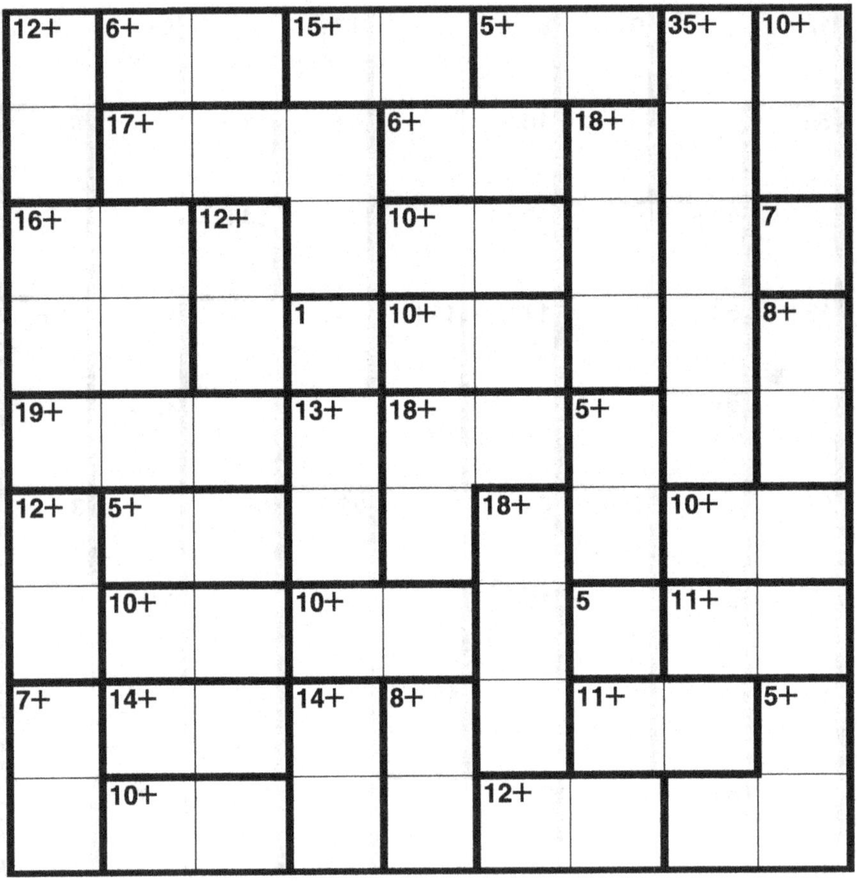

SOLVING
TIME

6

+

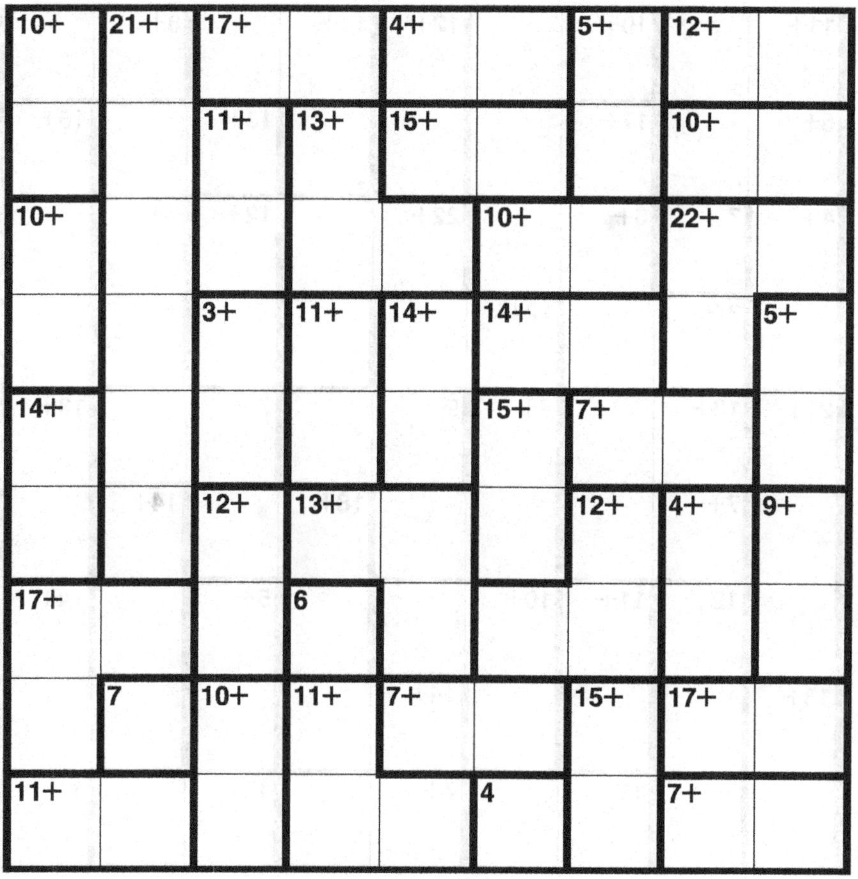

SOLVING
TIME

7

+

11+		10+		12+	11+		8+	
6+		17+				13+		16+
4+	7	8+		22+		12+	3+	
	8+		12+					
21+	13+		8+	9+		7+		13+
	7+				18+		14+	
	12+	11+	10+			5+		14+
13+				14+				
		11+		7+		17+		1

SOLVING
TIME

8

+

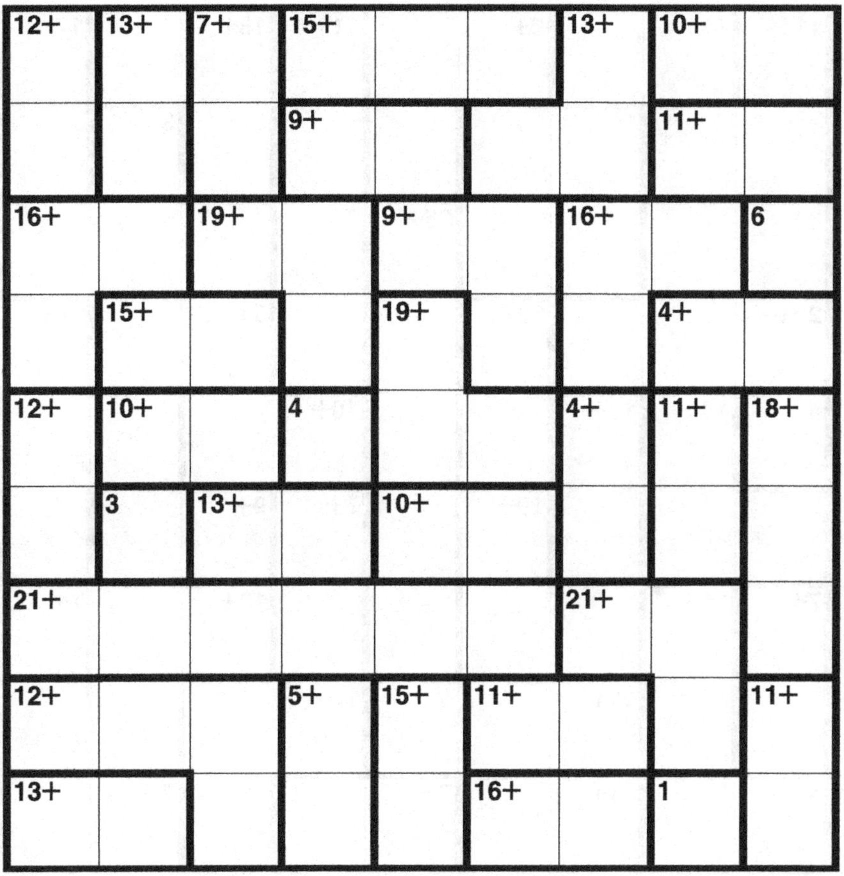

SOLVING
TIME

9

+

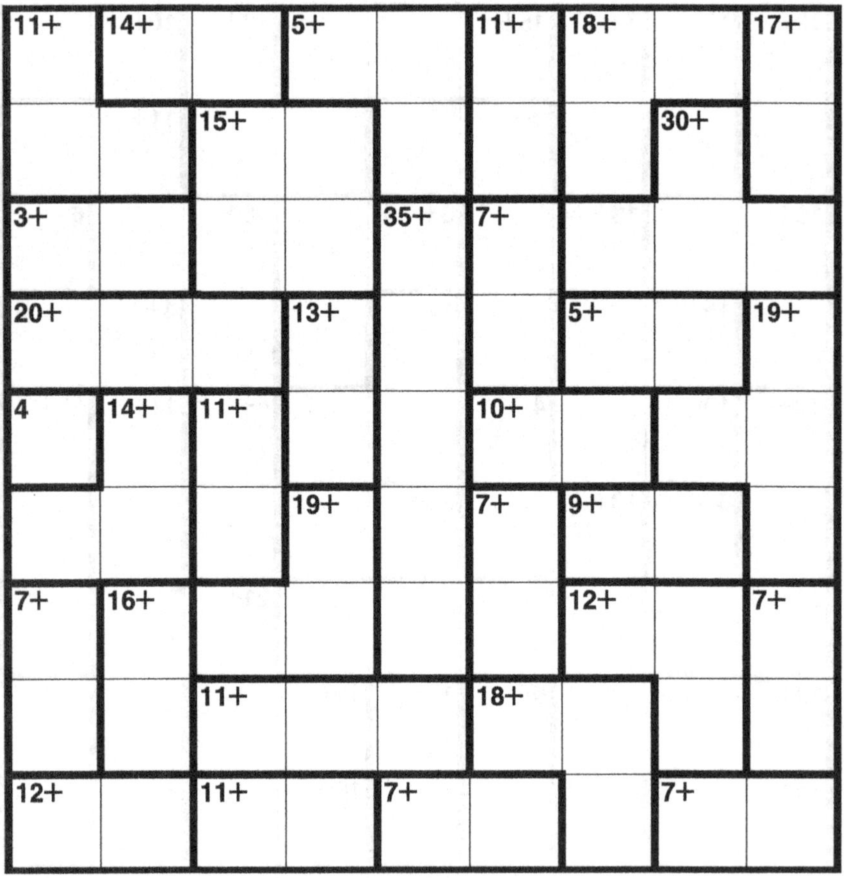

SOLVING
TIME

10

+

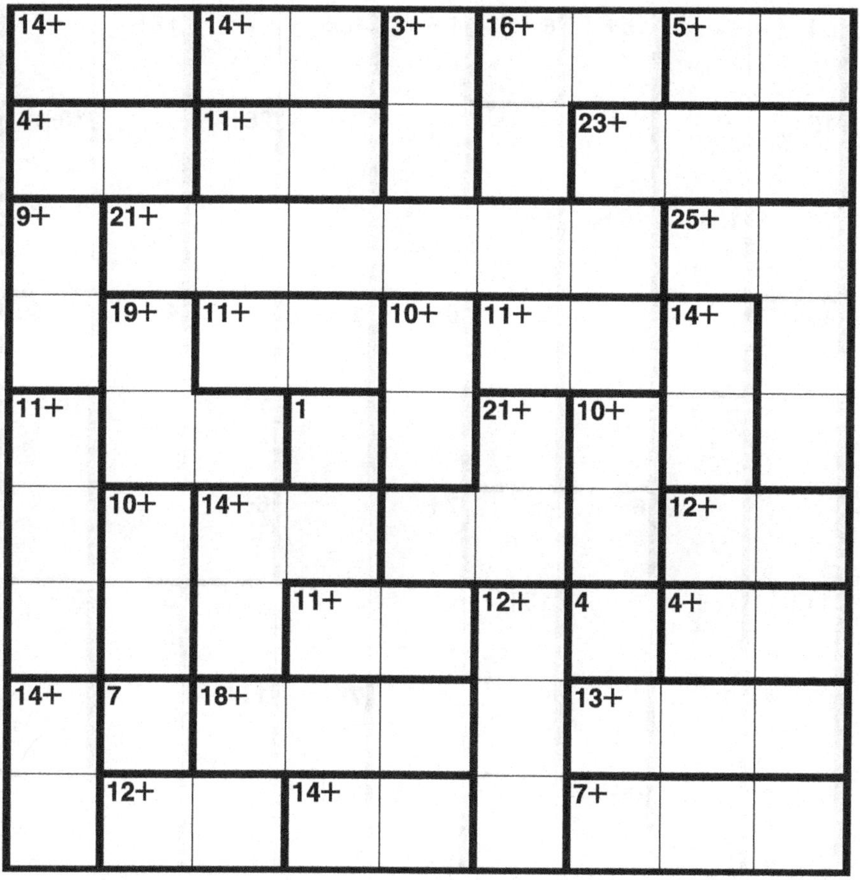

SOLVING
TIME

11

+ −

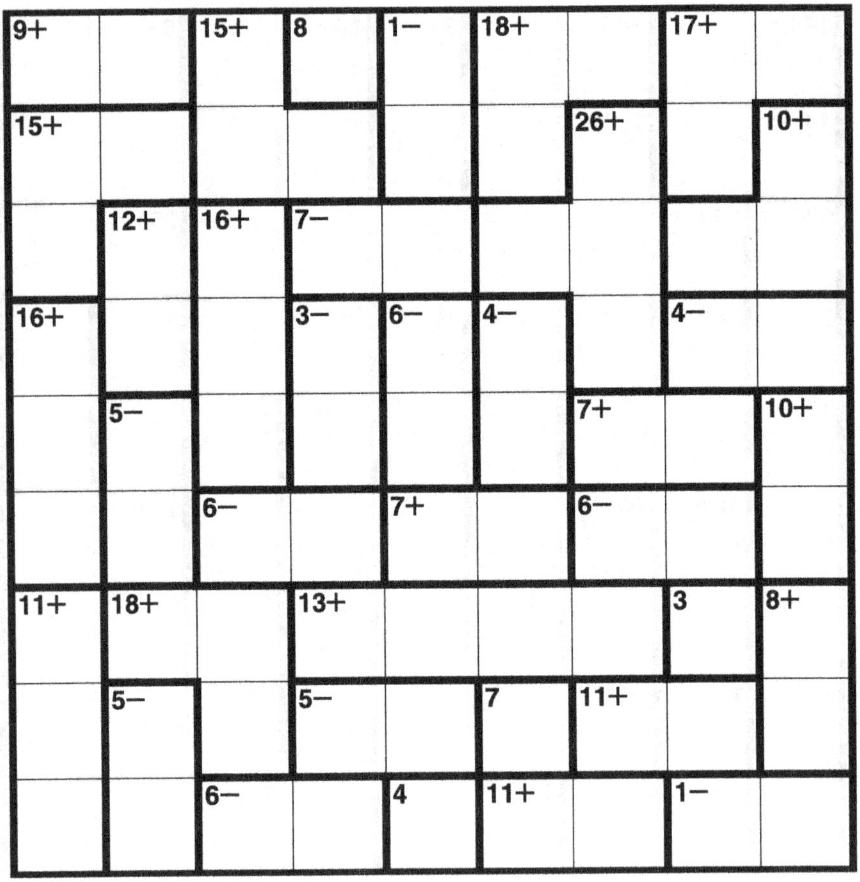

SOLVING
TIME

12

\+ −

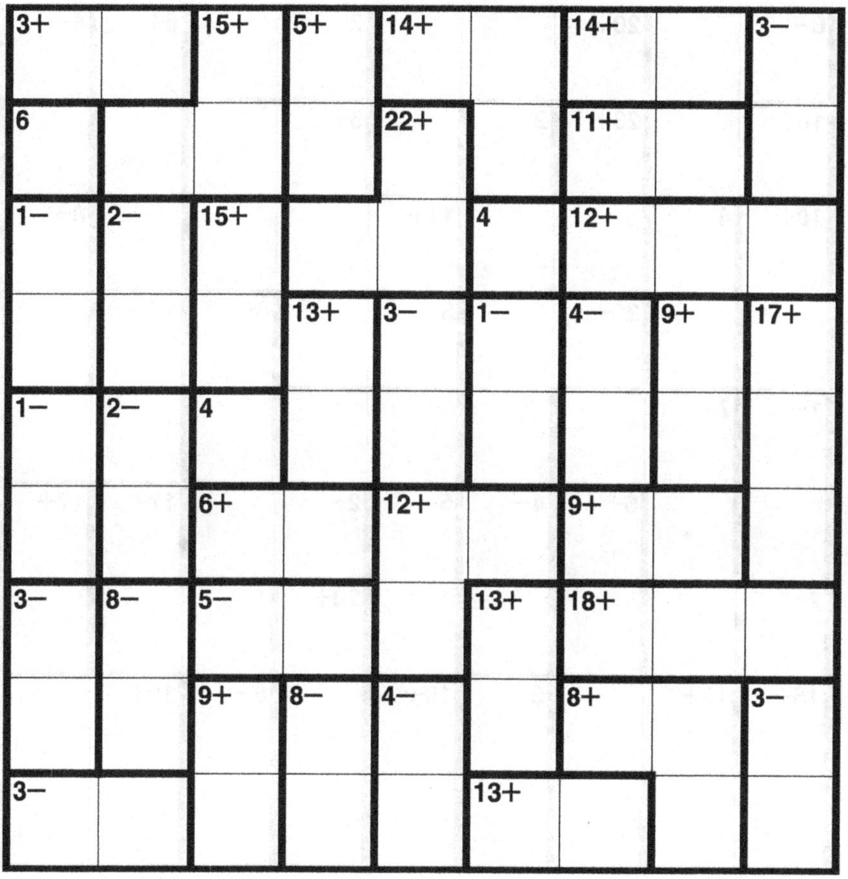

SOLVING
TIME

13

\+ −

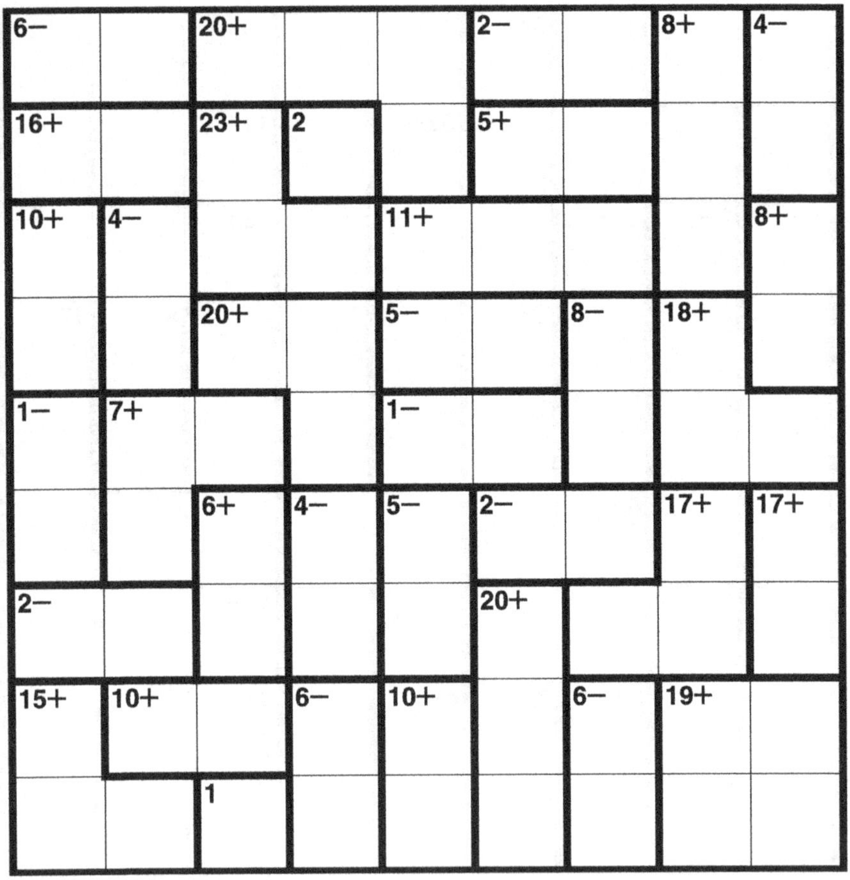

SOLVING
TIME

14

\+ −

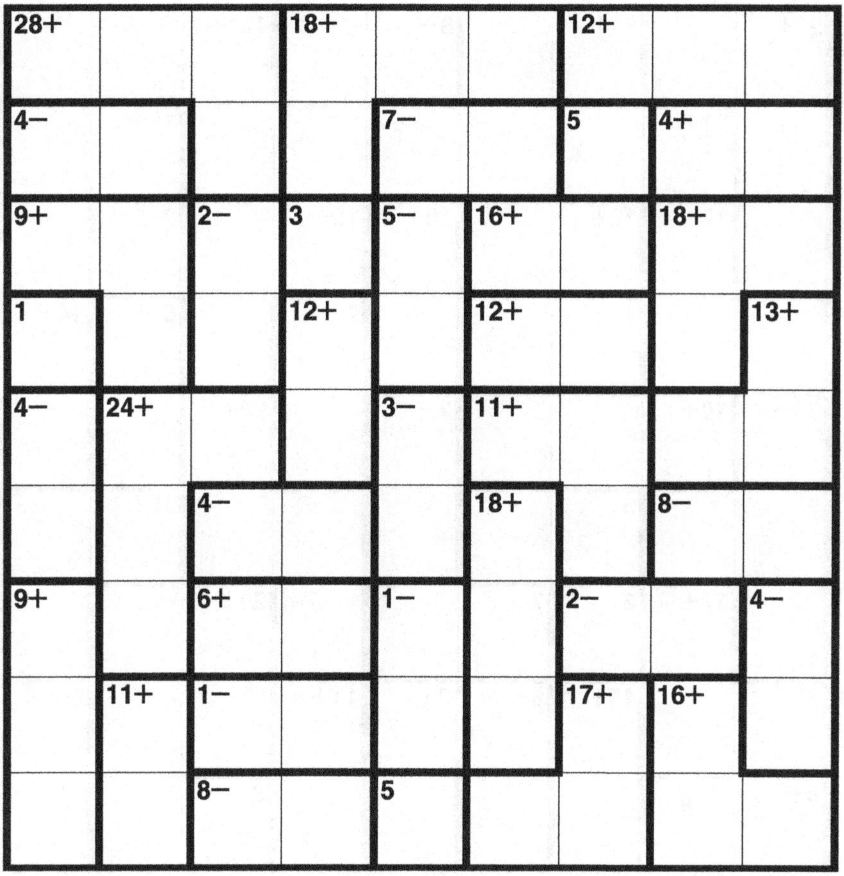

SOLVING
TIME

15

\+ −

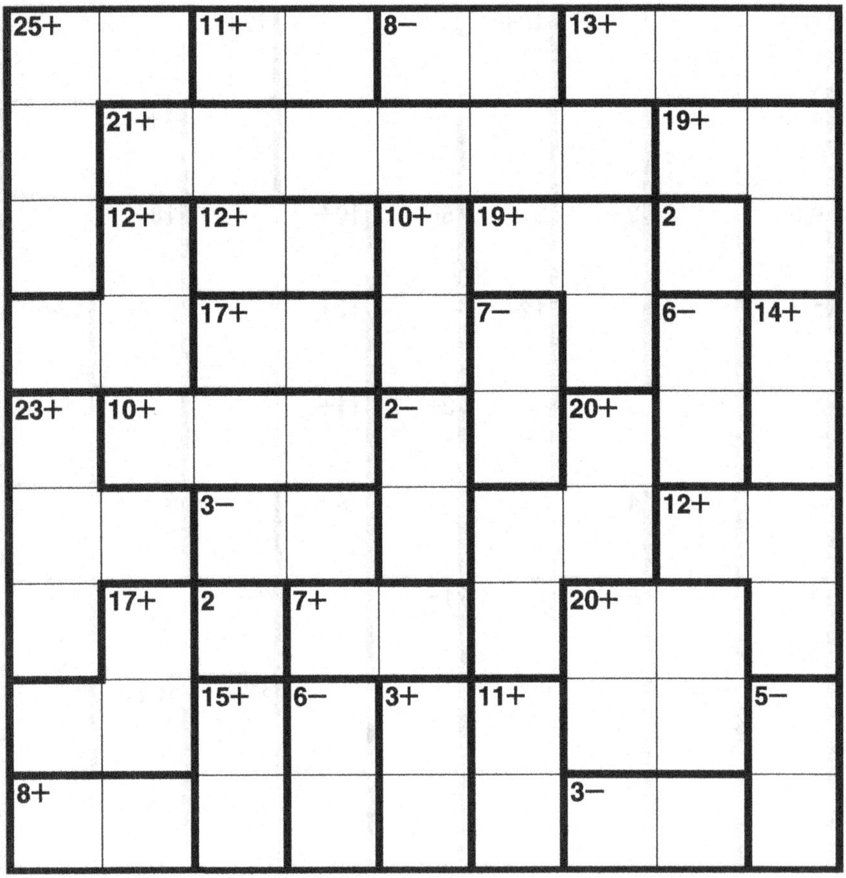

SOLVING

TIME

16

+ −

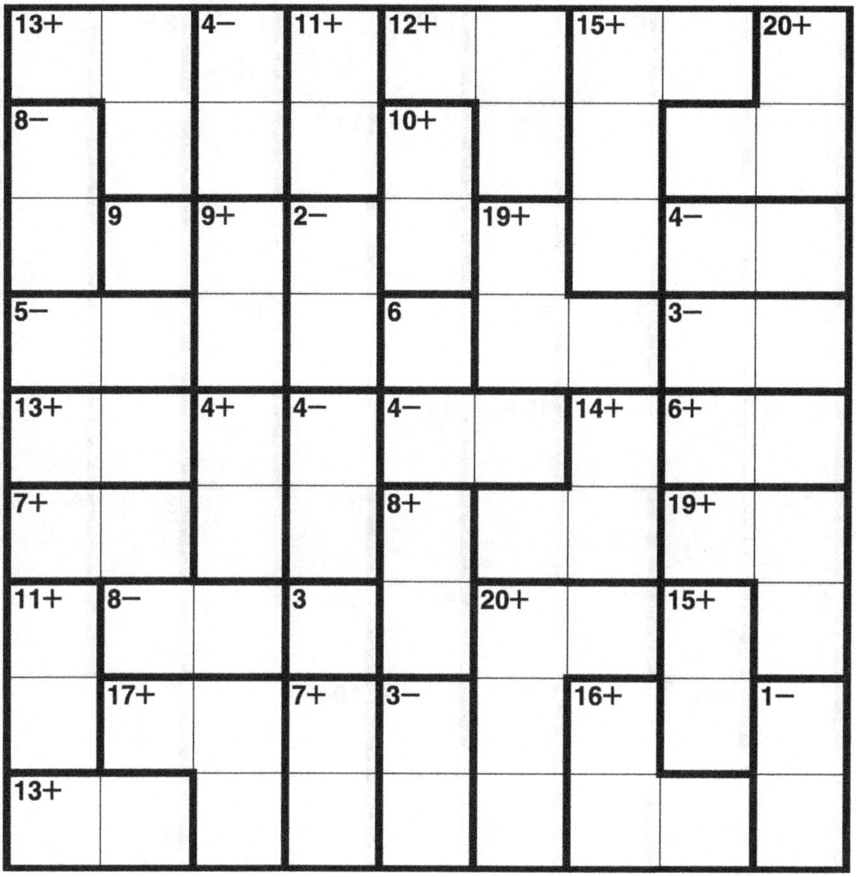

SOLVING
TIME

17

+ −

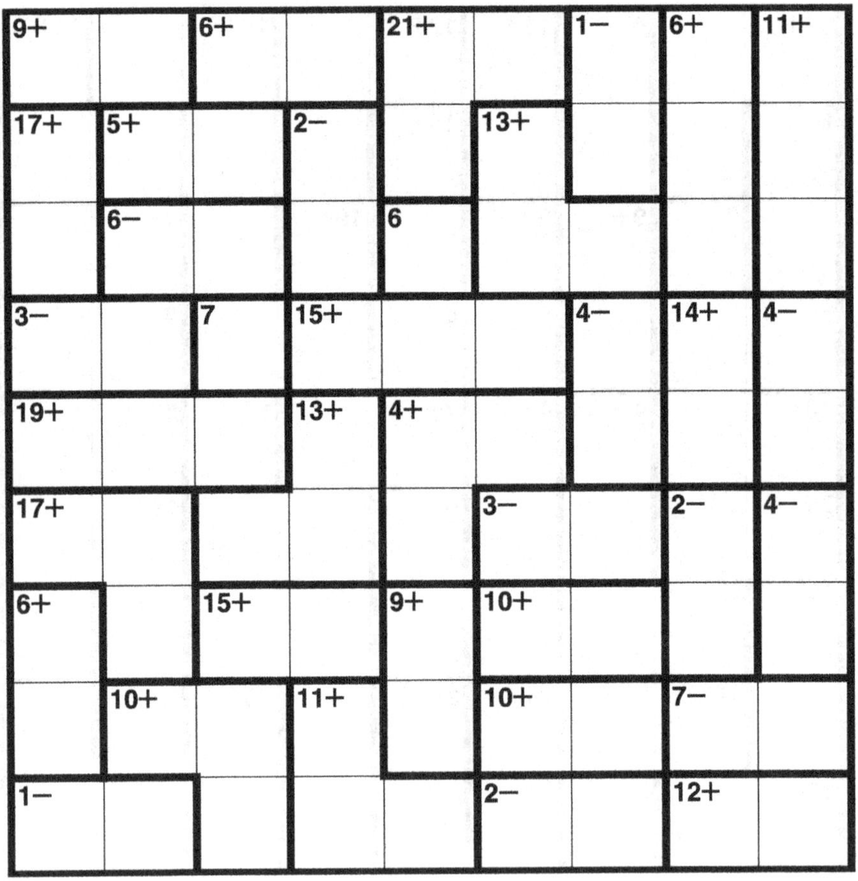

SOLVING TIME

18

+ −

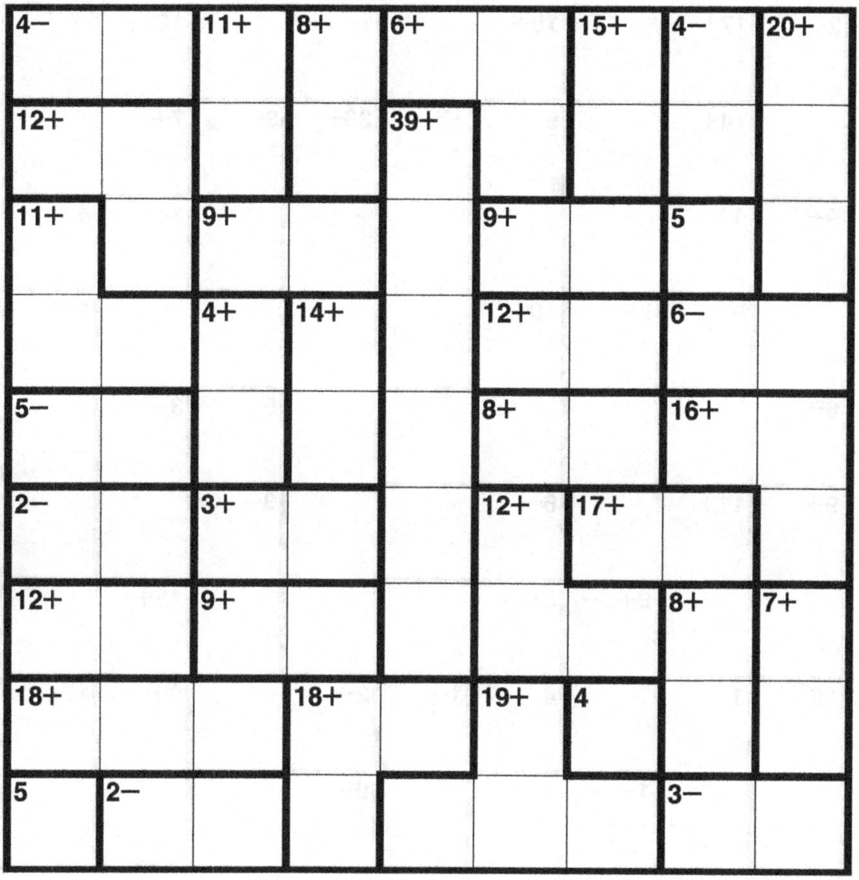

SOLVING
TIME

19

\+ −

2−	17+		10+		1−		12+	
	14+		9+	17+	39+	3−	7+	
9+	4+						14+	4
	9	9+	18+					11+
6−						5	3−	
9+	11+		6−			3−		
		9+	2−				15+	
15+	1−		4	1−	2−		9+	17+
		1−			9+			

SOLVING
TIME

20

+ −

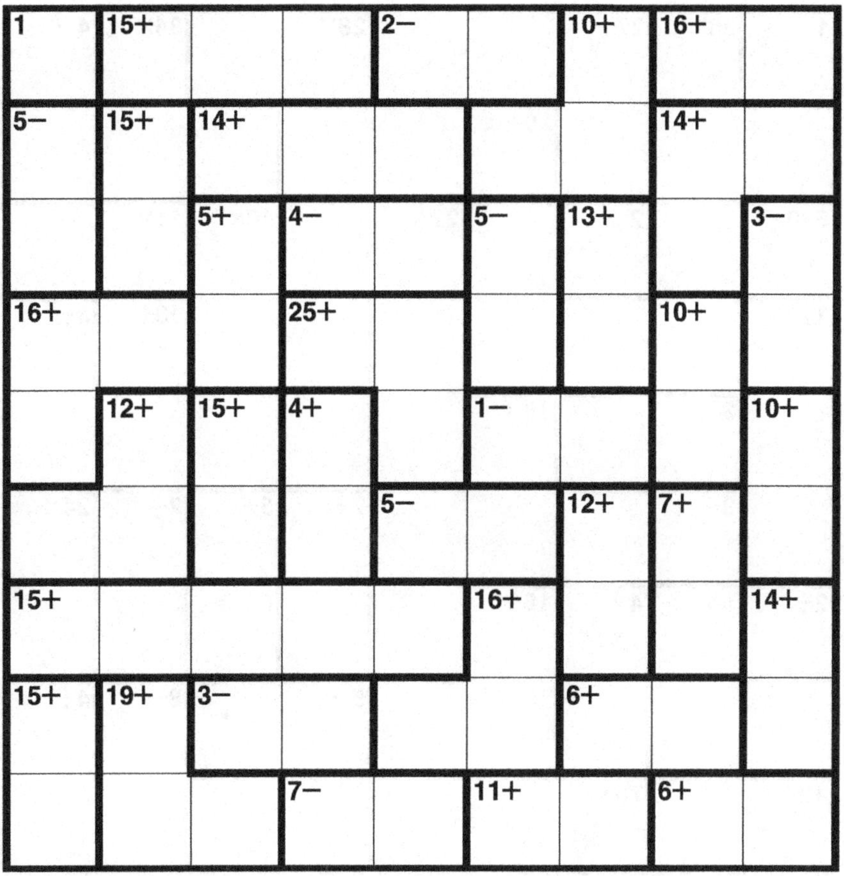

SOLVING
TIME

21

+ − × ÷

1	30×	27+			28×		84×	4−
			20+					
540×		2		22+		40×	21×	
112×				1−			13+	3÷
	3−		18×					
	3−				7+	5+	2÷	24×
2÷	56×	4−	18+					
					5−		8−	4÷
10+		70×			3÷			

SOLVING
TIME

22

+ − × ÷

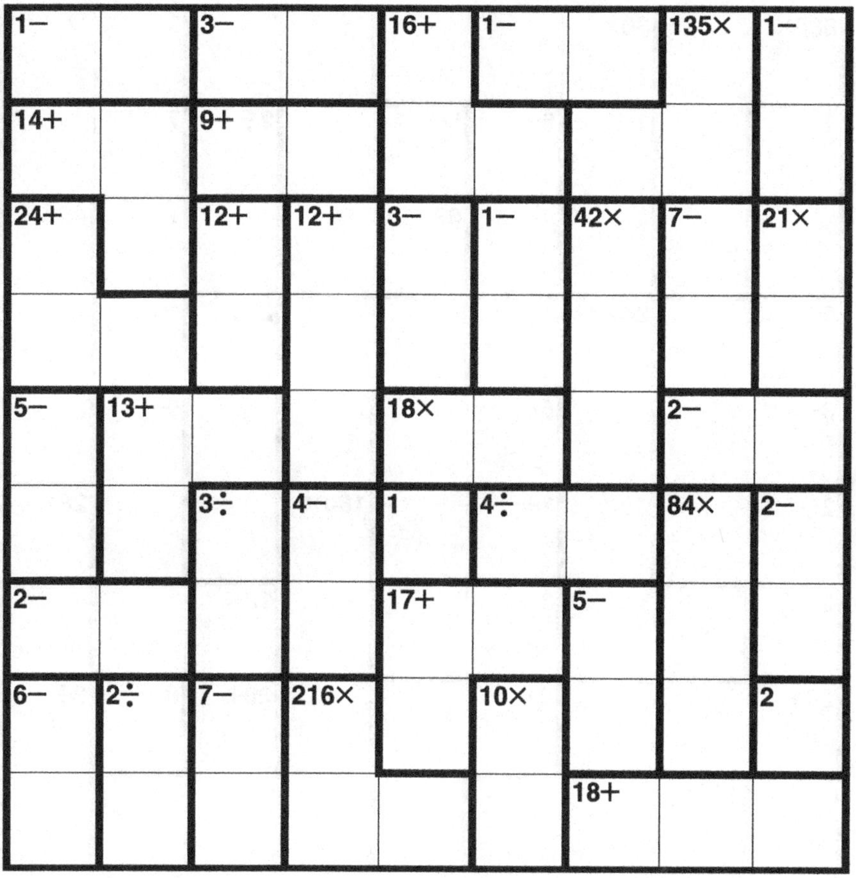

SOLVING
TIME

23

+ − × ÷

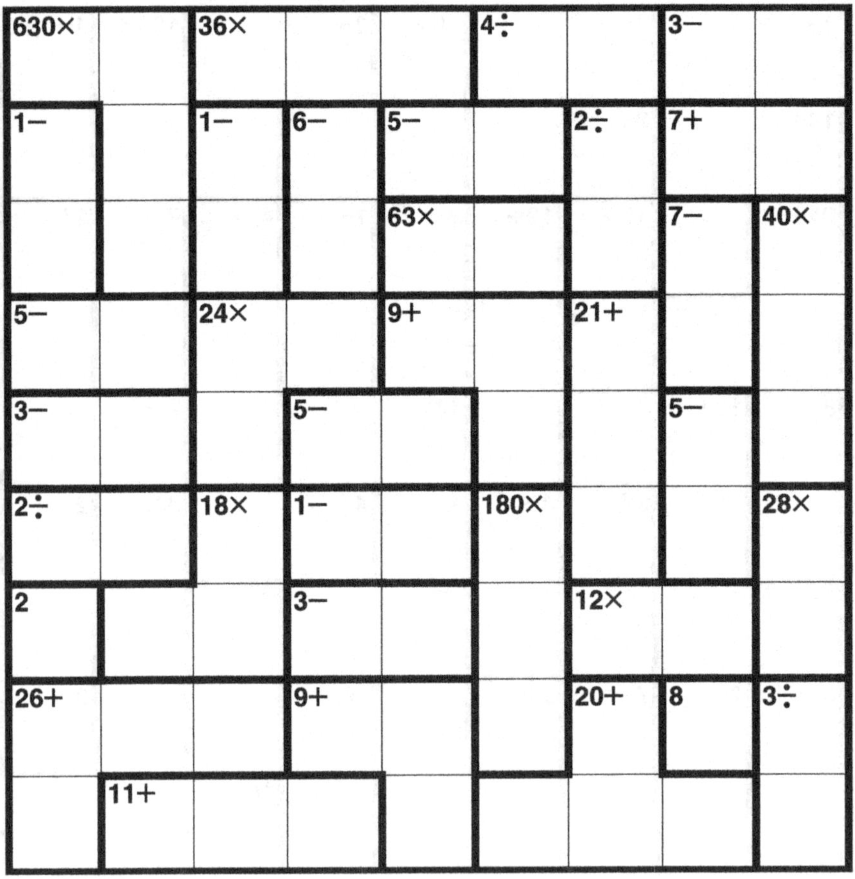

SOLVING
TIME

24

+ − × ÷

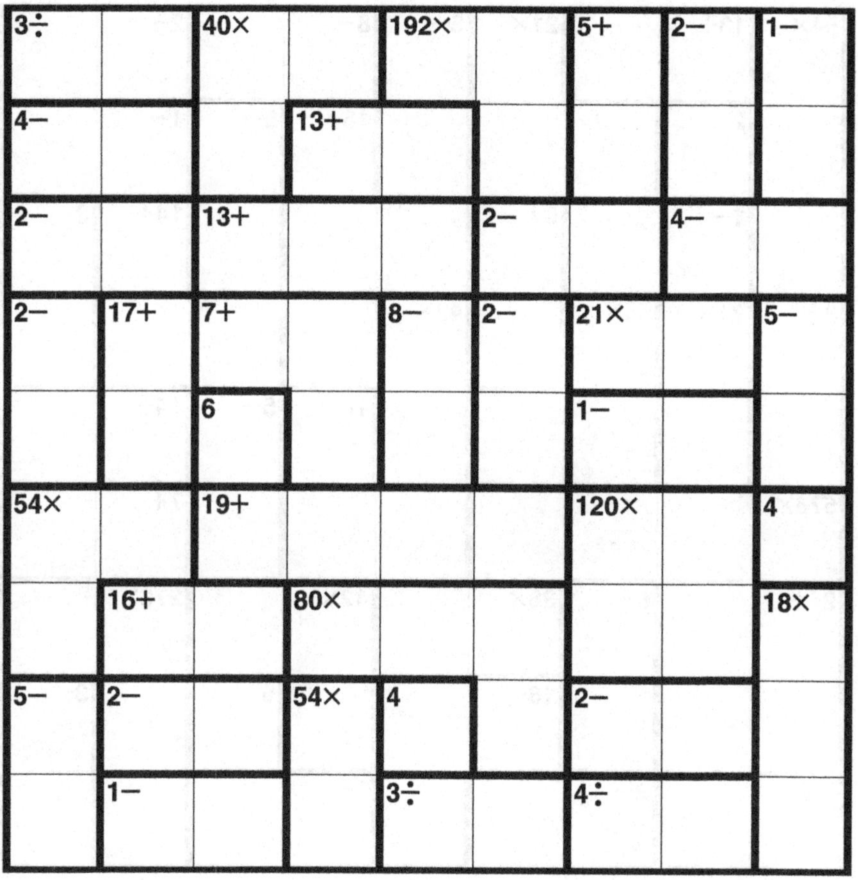

SOLVING
TIME

25

+ − × ÷

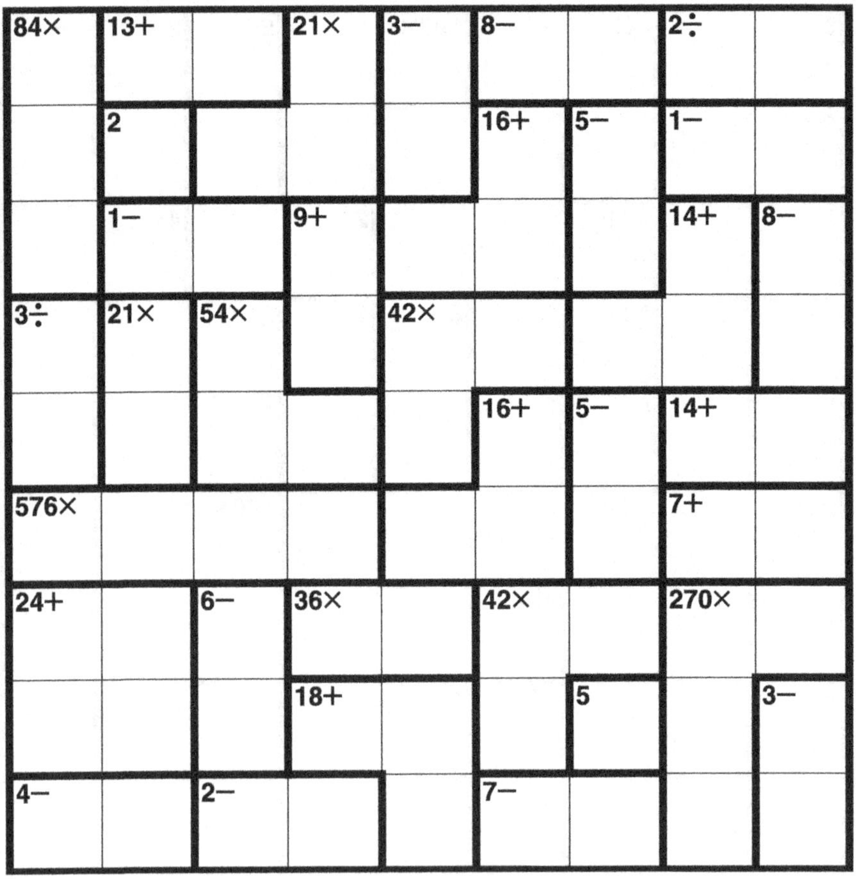

SOLVING
TIME

26

+ − × ÷

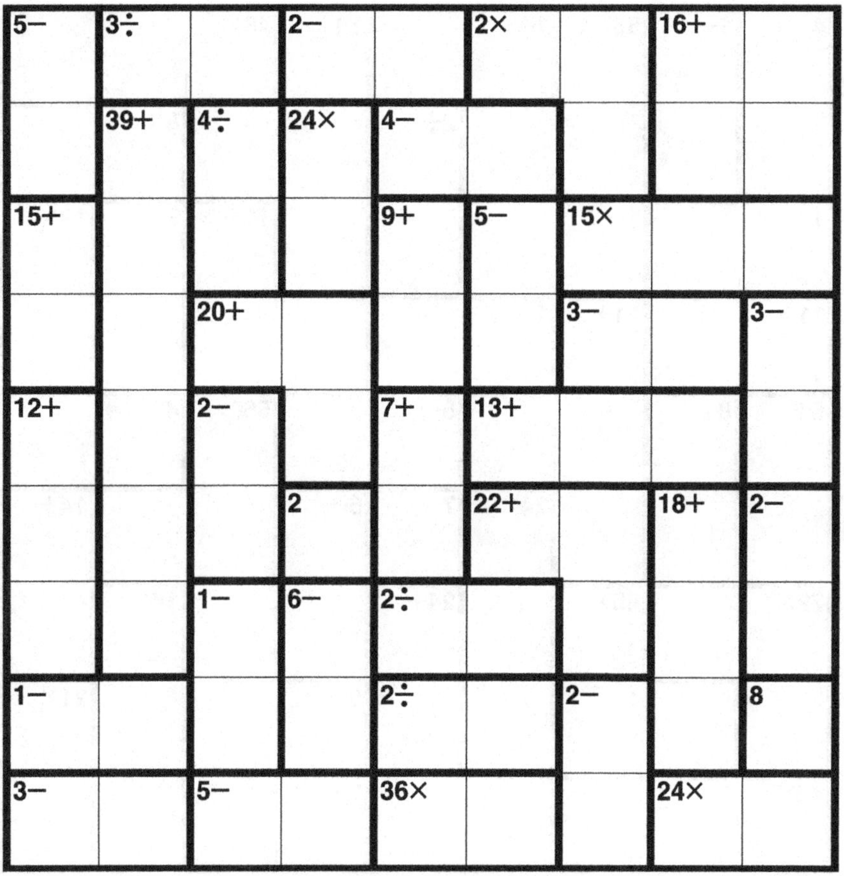

SOLVING
TIME

27

+ − × ÷

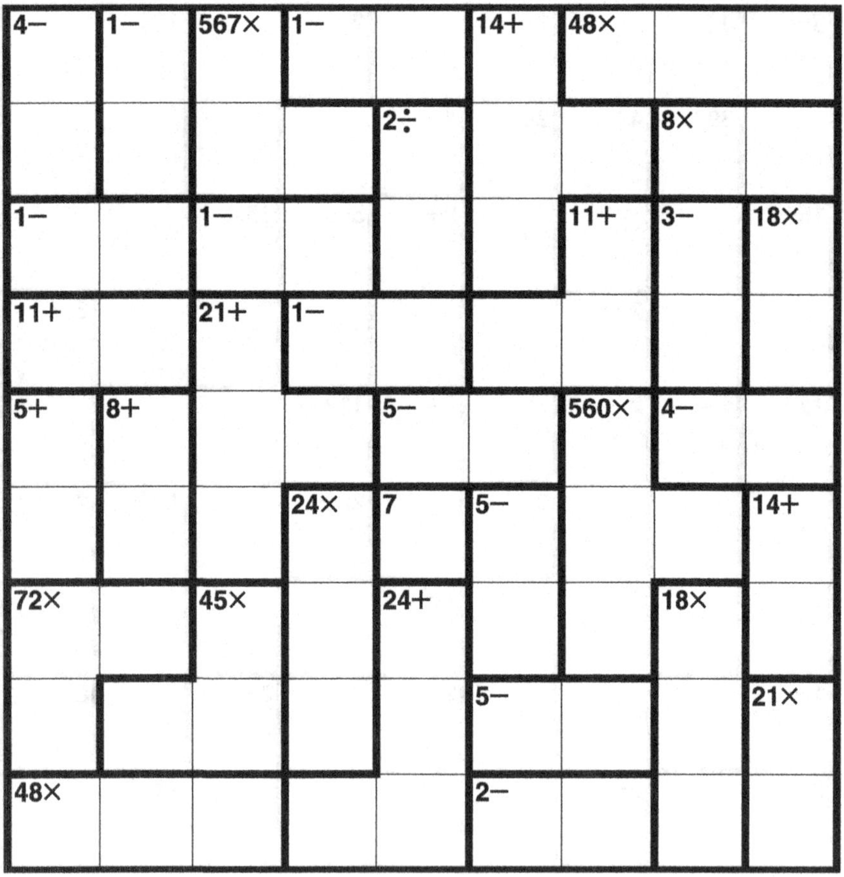

SOLVING
TIME

28

+ − × ÷

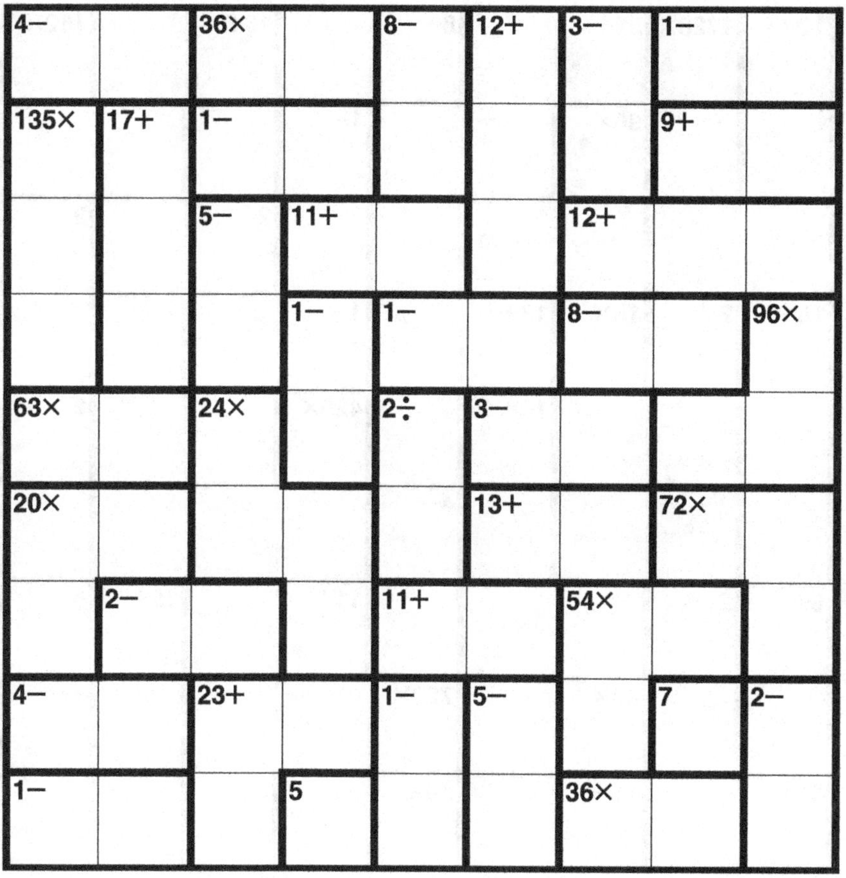

SOLVING
TIME

29

\+ − × ÷

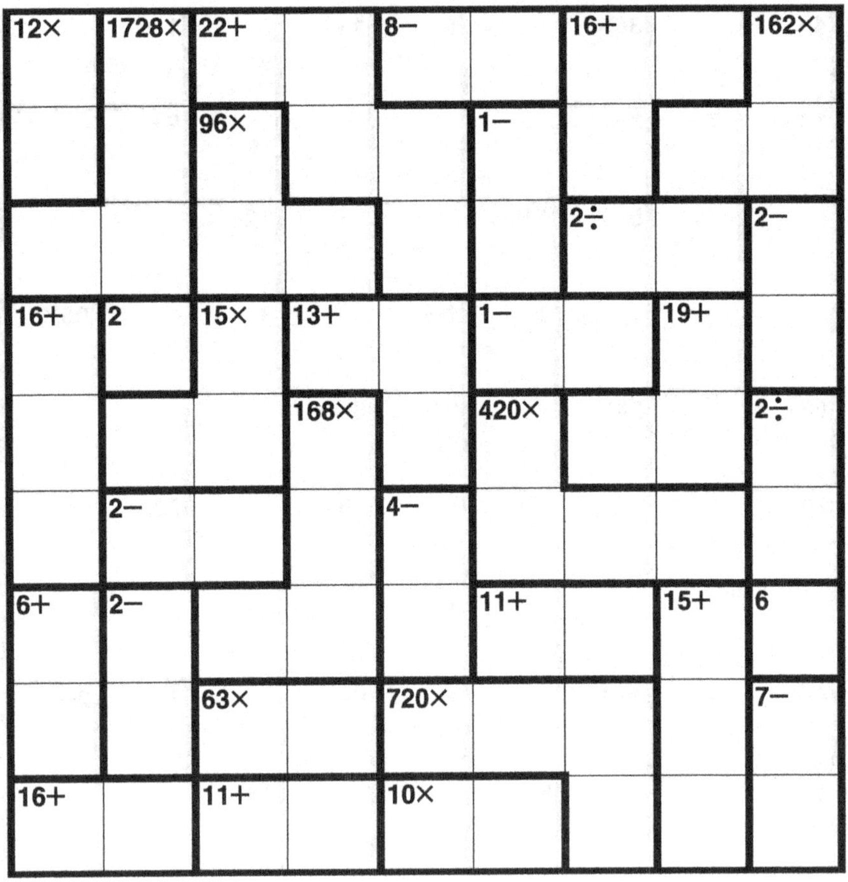

SOLVING TIME

30

+ − × ÷

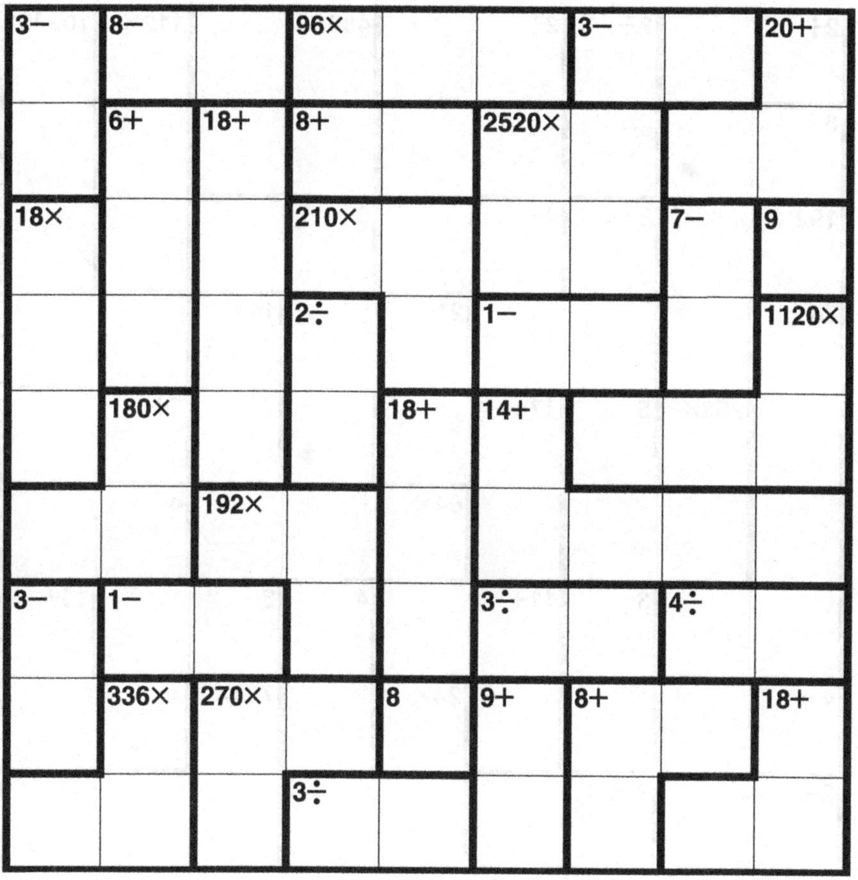

SOLVING
TIME

31

+ − × ÷

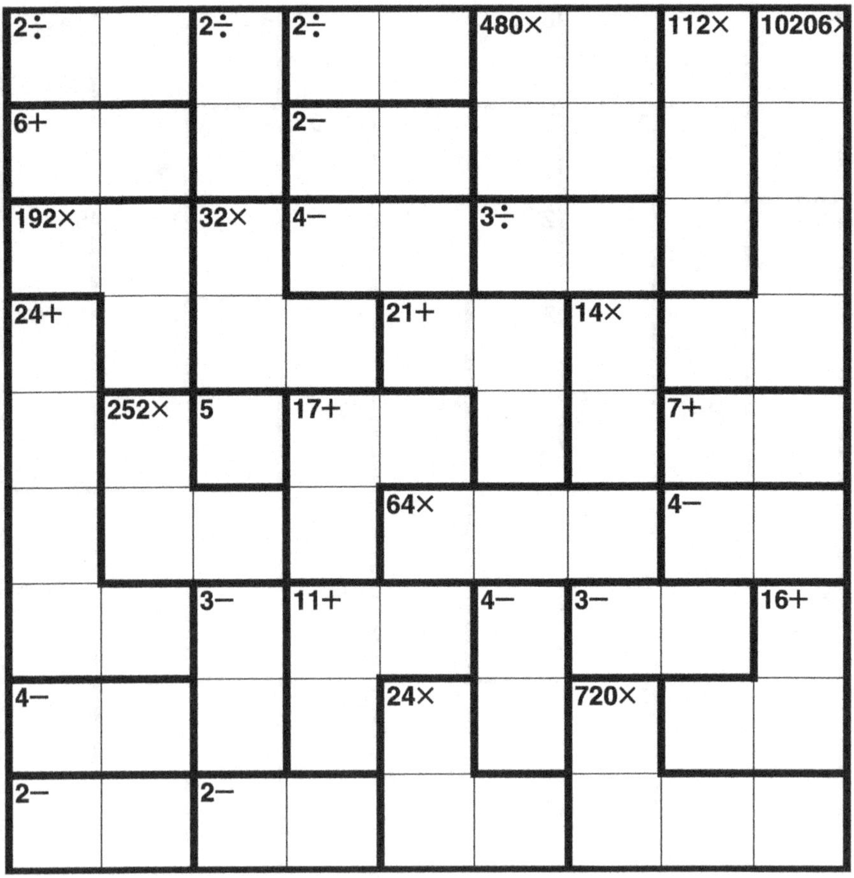

SOLVING
TIME

32

+ − × ÷

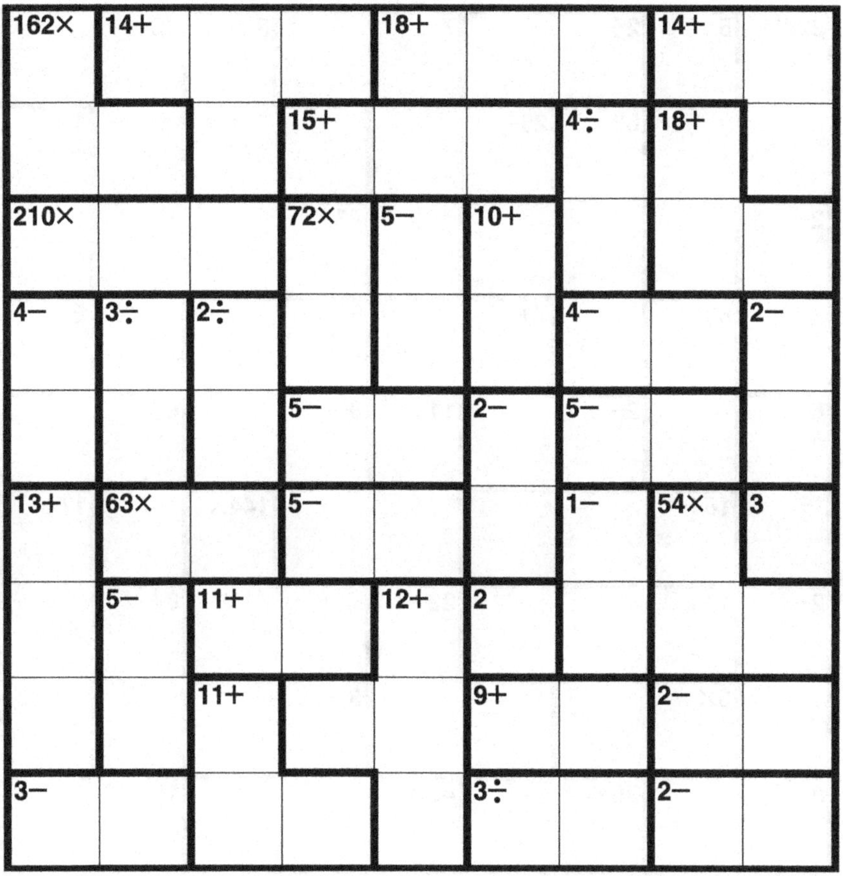

SOLVING
TIME

33

+ − × ÷

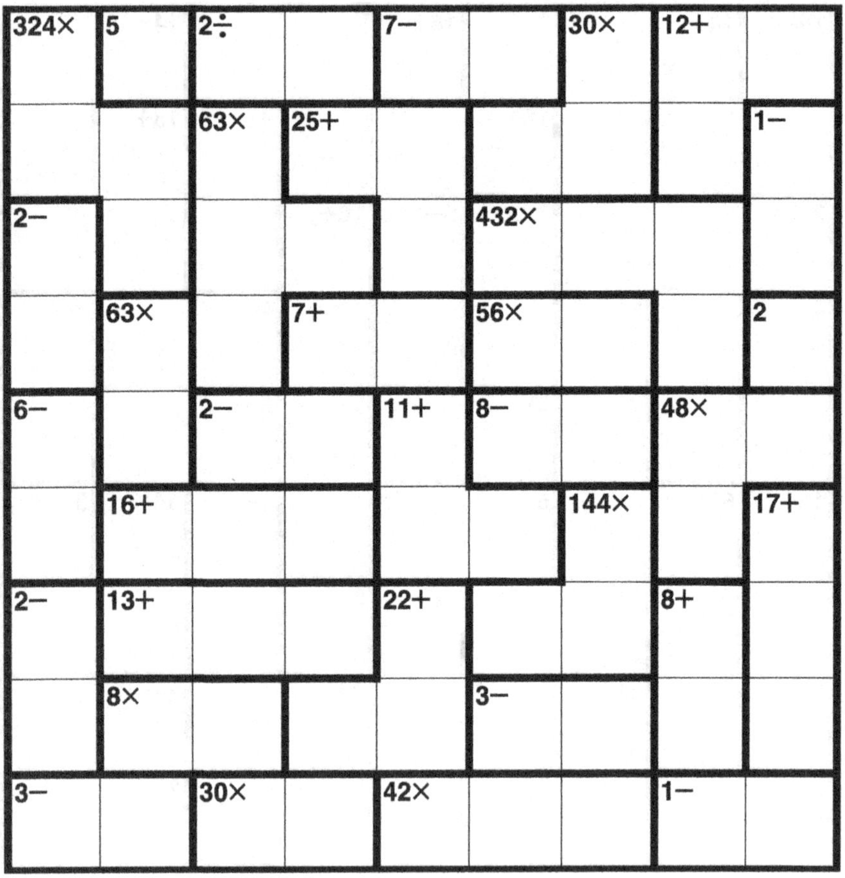

SOLVING
TIME

34

+ − × ÷

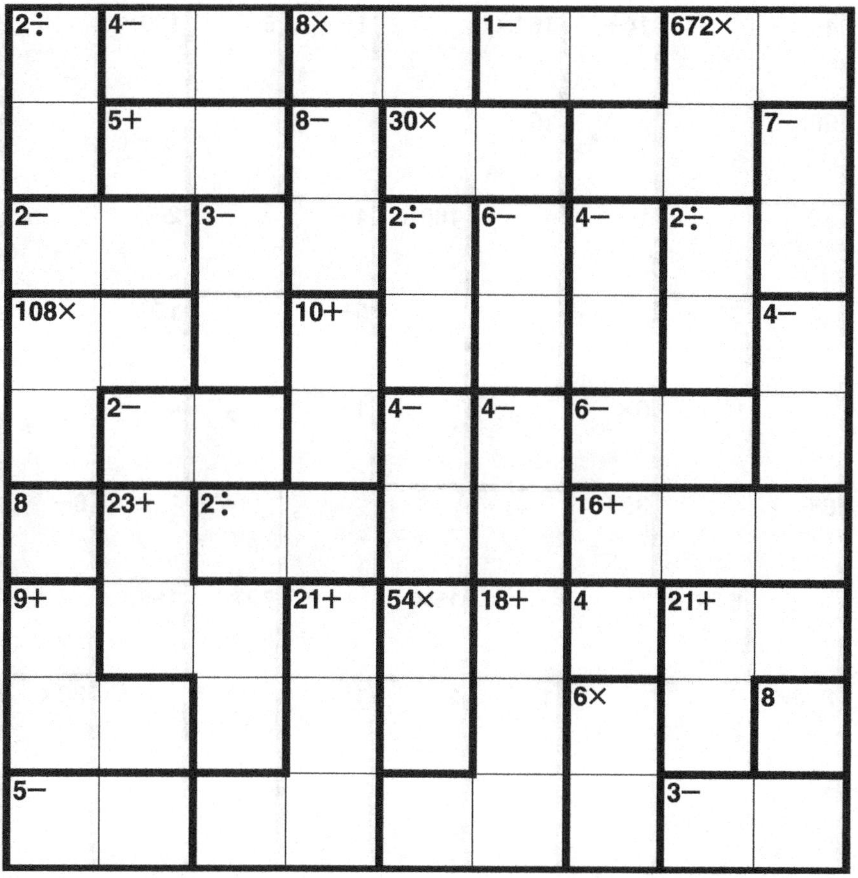

SOLVING
TIME

35

+ − × ÷

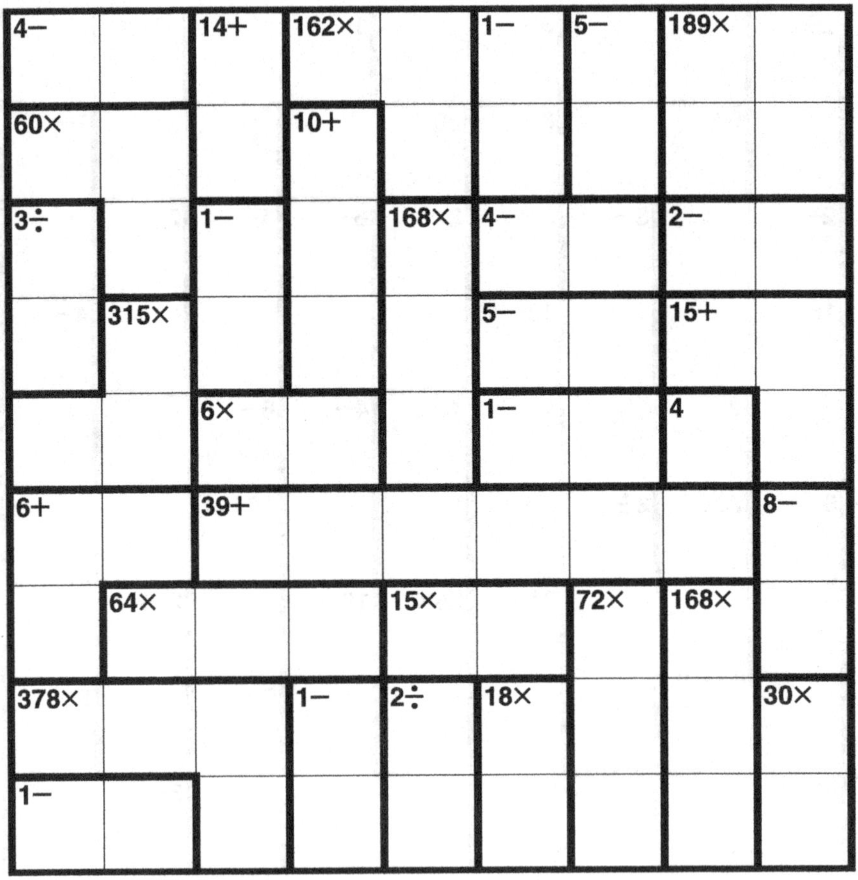

SOLVING
TIME

36

+ − × ÷

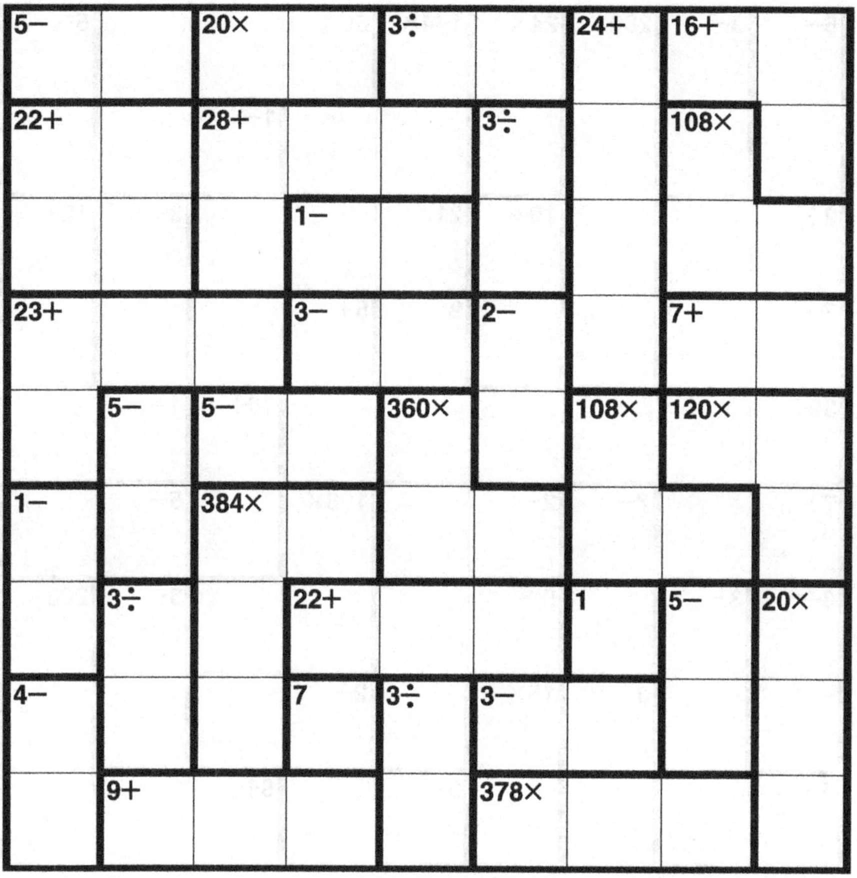

SOLVING
TIME

37

+ − × ÷

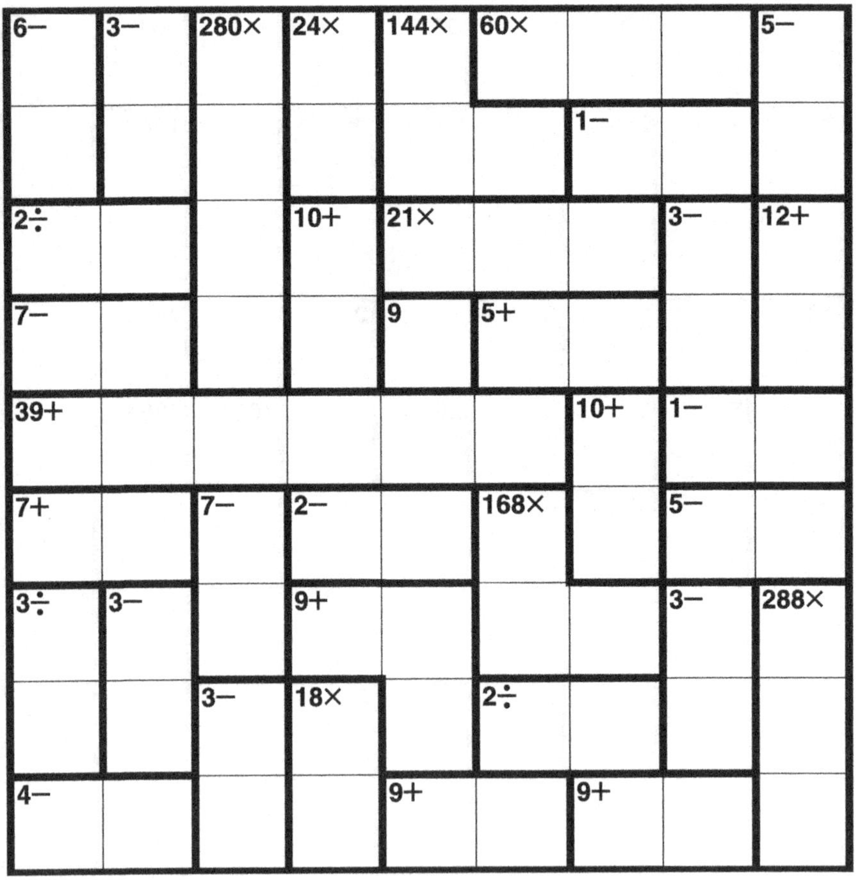

SOLVING
TIME

38

+ − × ÷

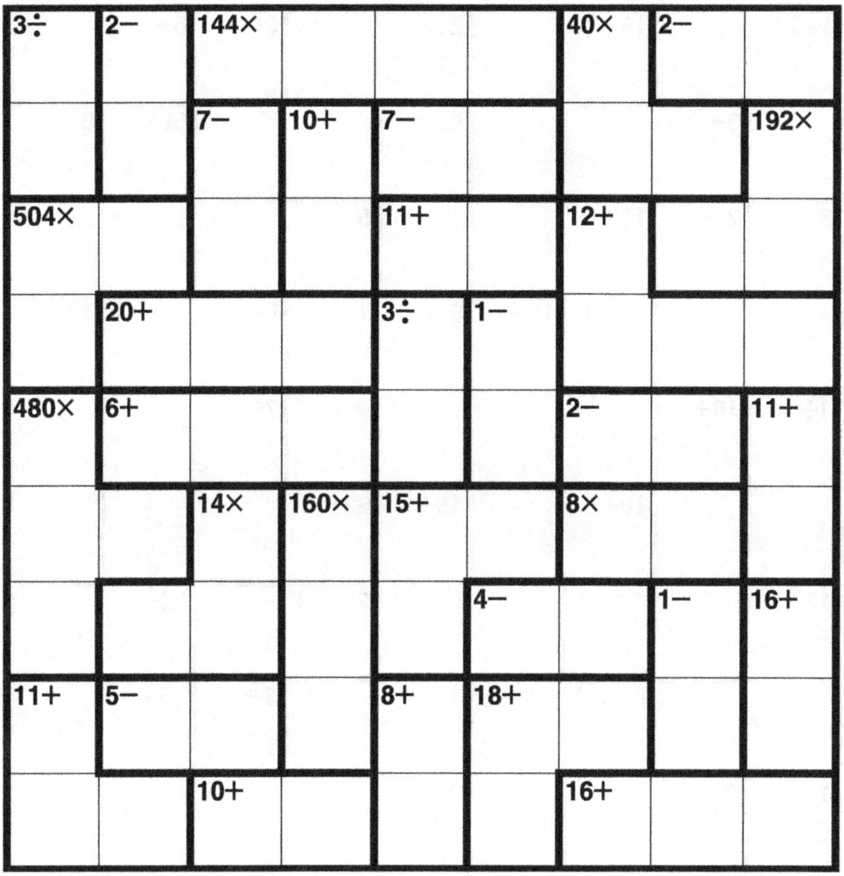

SOLVING

TIME

39

+ − × ÷

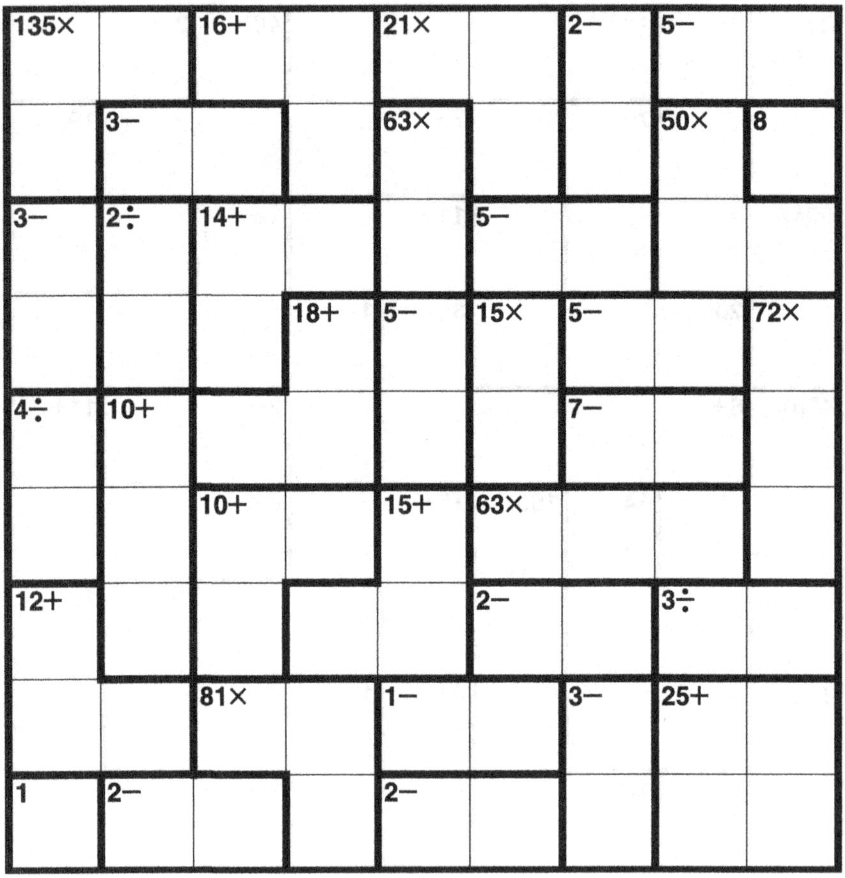

SOLVING
TIME

40

+ − × ÷

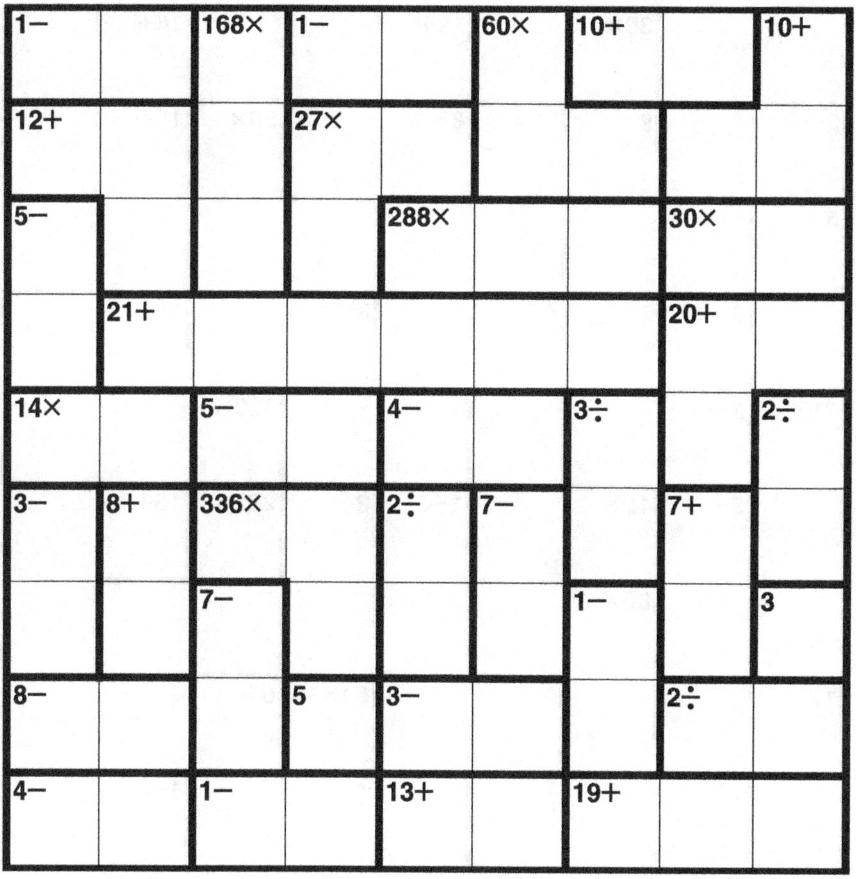

SOLVING
TIME

41

+ − × ÷

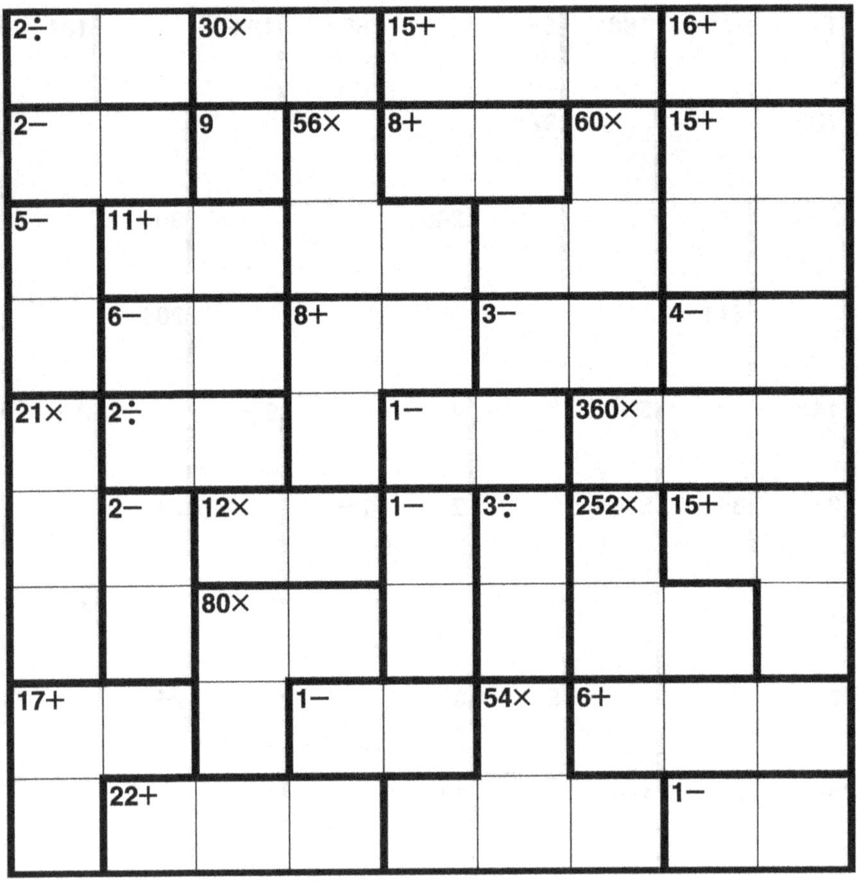

SOLVING
TIME

42

+ − × ÷

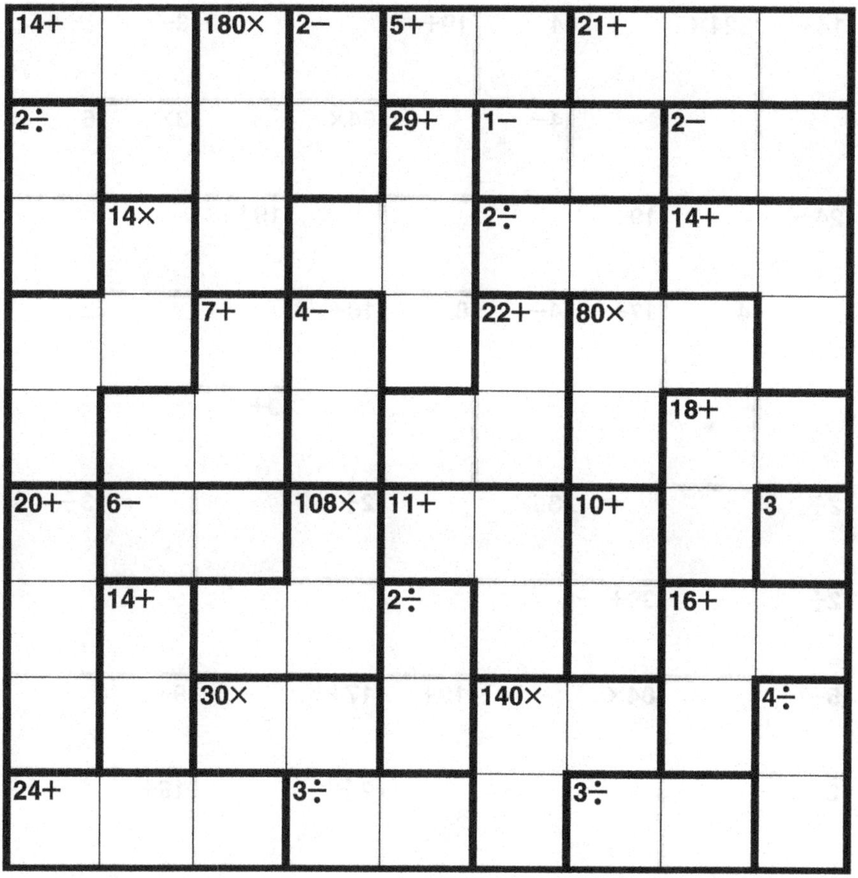

SOLVING
TIME

43

+ − × ÷

14+	324×		4	10+			3−	
			4−		64×		3×	6
24+		19+				19+		
	4	17+	4−	8−	10+			2−
						9+		
2÷			35×		2÷			3÷
2÷		39+						
5−		84×		12+	17+		9+	
2−					2÷		16+	

SOLVING TIME

44

+ − × ÷

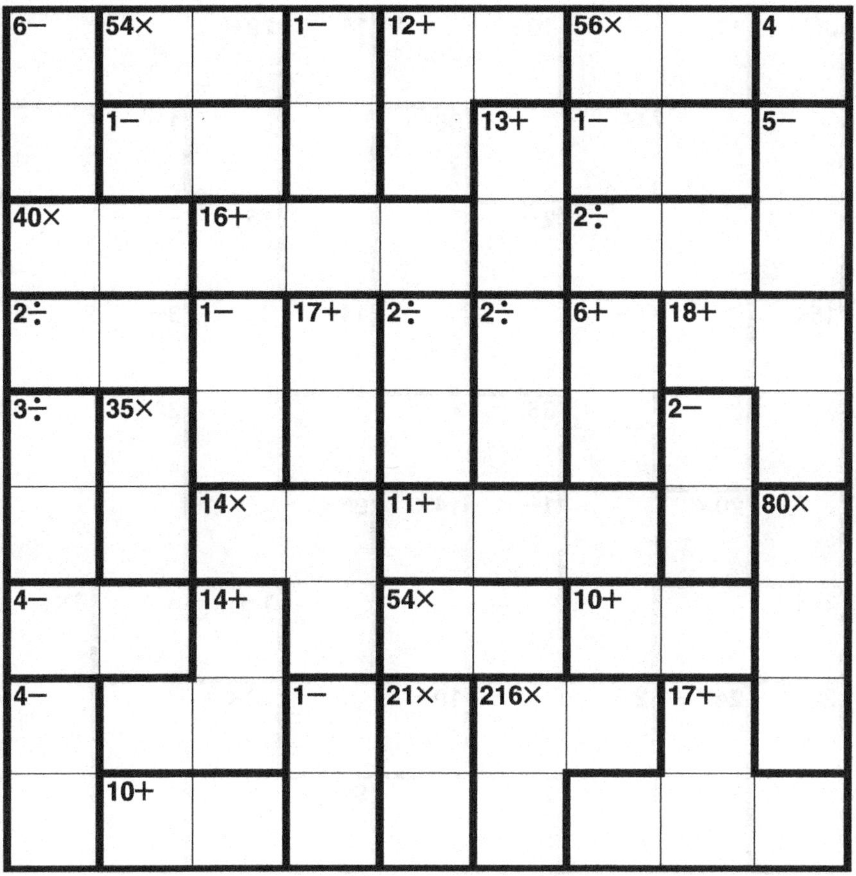

SOLVING
TIME

45

+ − × ÷

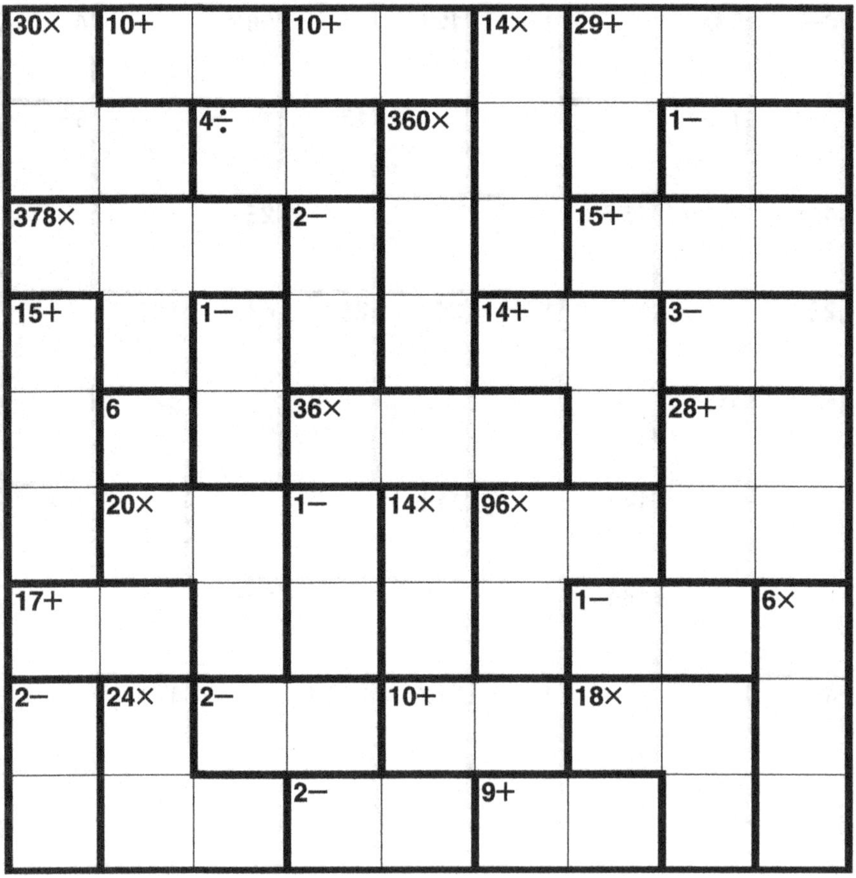

SOLVING
TIME

46

+ − × ÷

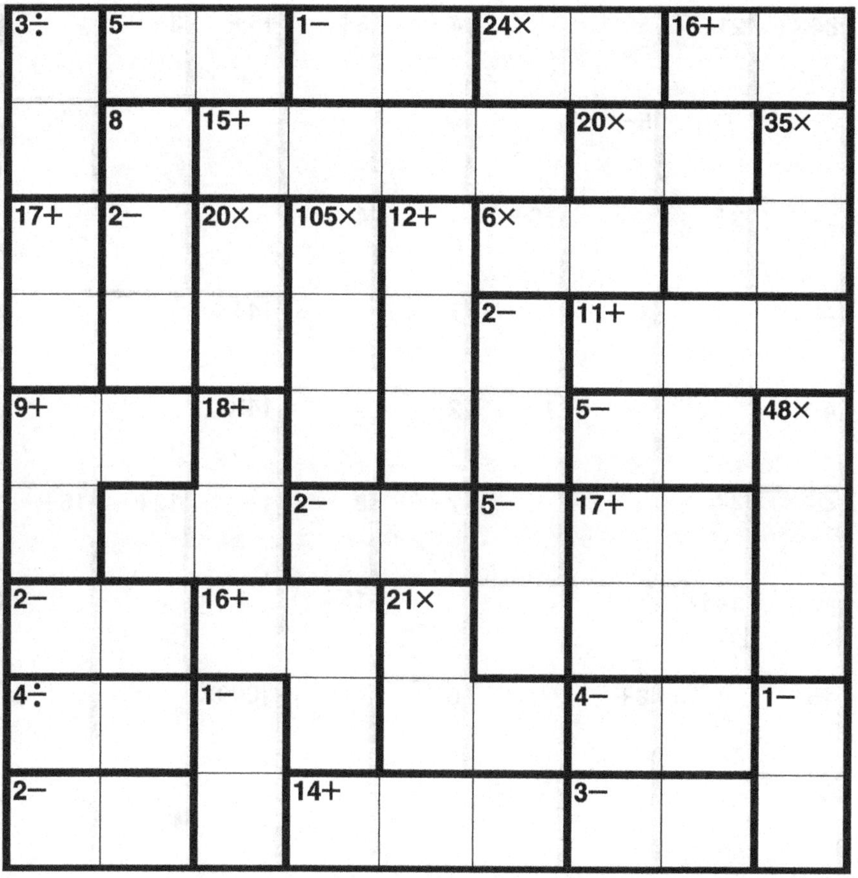

SOLVING
TIME

47

+ − × ÷

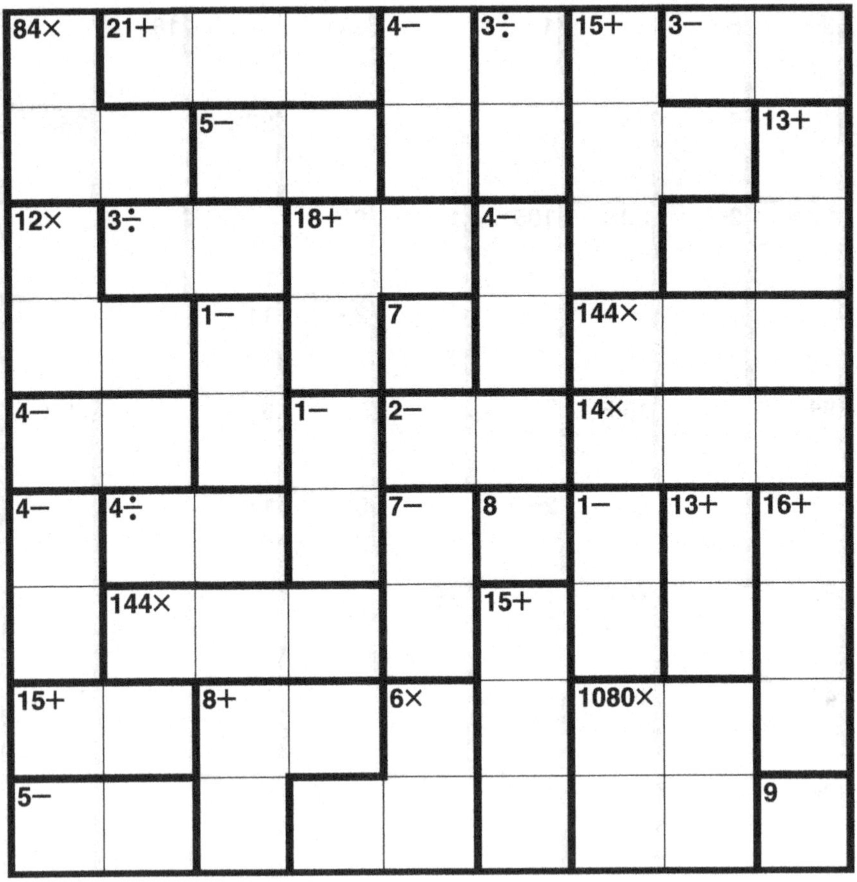

SOLVING
TIME

48

+ − × ÷

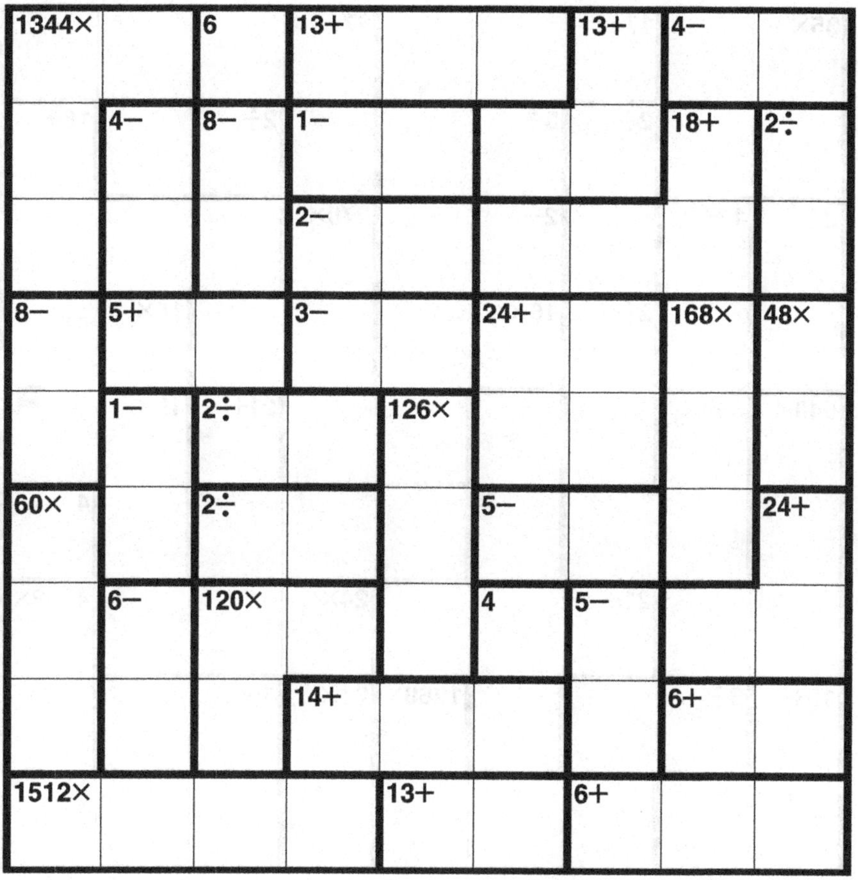

SOLVING
TIME

49

+ − × ÷

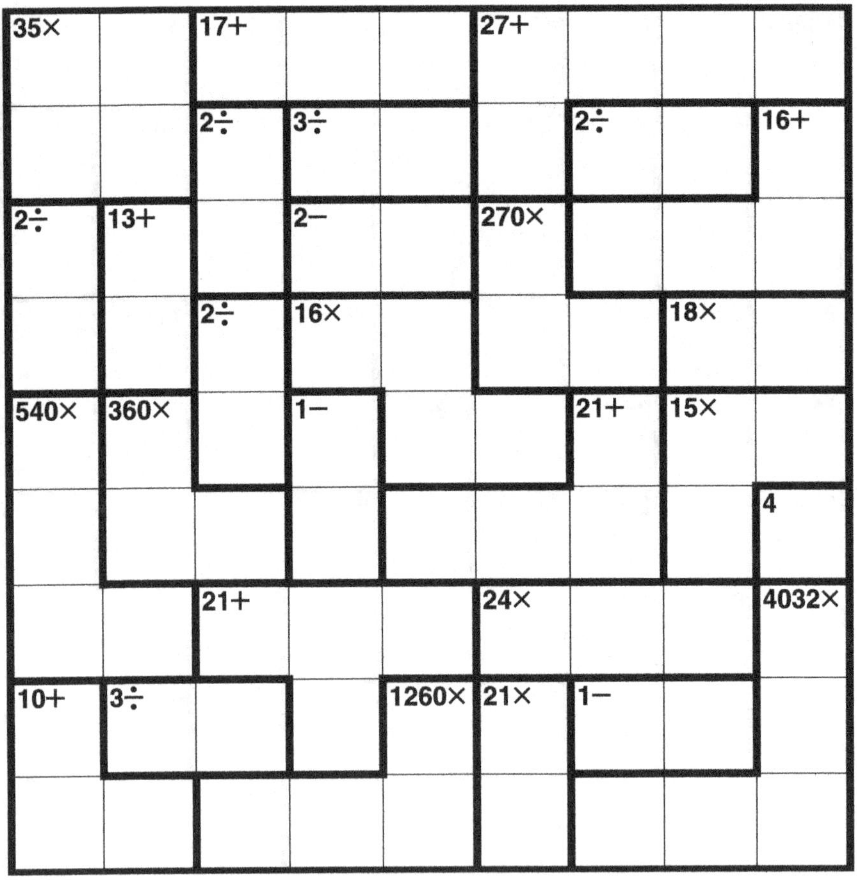

SOLVING
TIME

50

+ − × ÷

30×	26+		1−		21+	12×	28+	
			10+				17+	
504×		20×		3−		18×		
6−		2÷						14+
	3÷	22+				1−		
1−		42×	40×				17+	
	2÷			5−	504×	40×		
19+		1−					15×	6+
		7	4÷					

SOLVING
TIME

51

\+ − × ÷

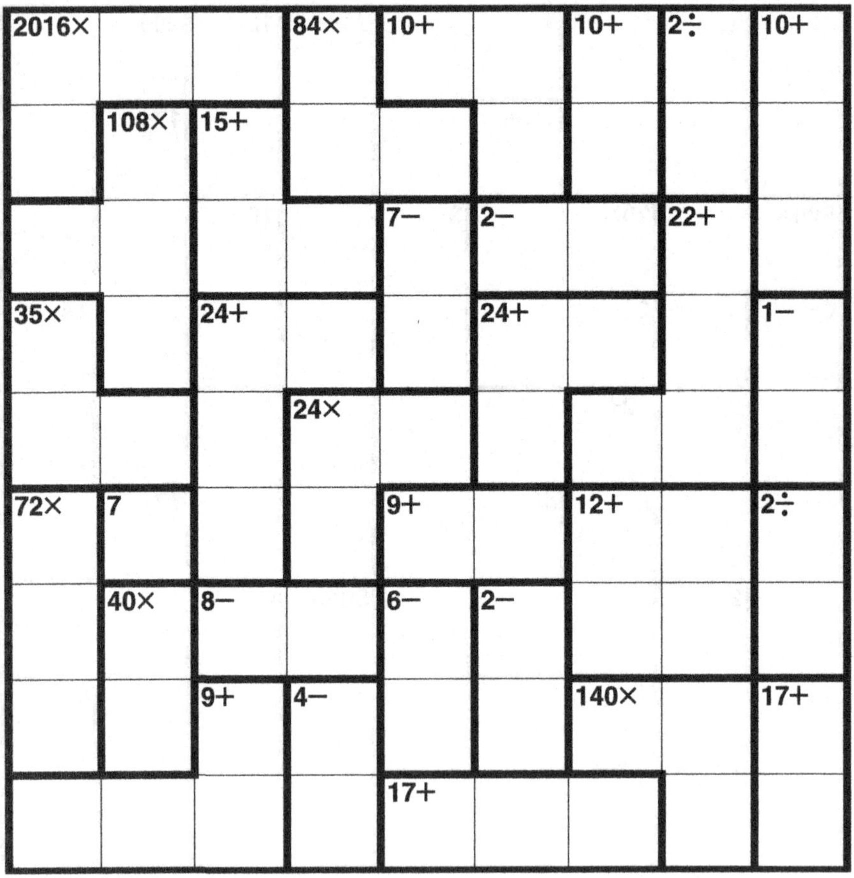

SOLVING
TIME

52

+ − × ÷

8×		3÷	8+		22+			12+
2−			3÷		12+	15+	4÷	
3−		63×						
48×	4	27×		10×			3−	
	3÷		28×		13+	24×	16+	
		1344×		2−				10+
4−					2		72×	
	12+			18+				5−
2−		7+		21+				

SOLVING
TIME

53

+ − × ÷

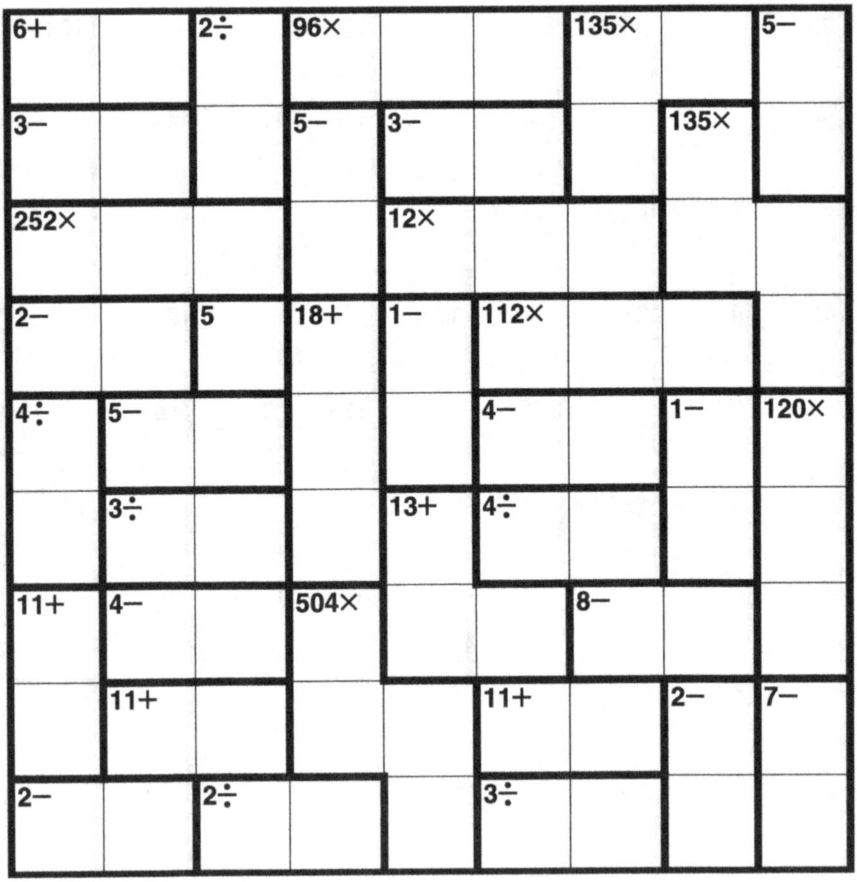

SOLVING TIME

54

+ − × ÷

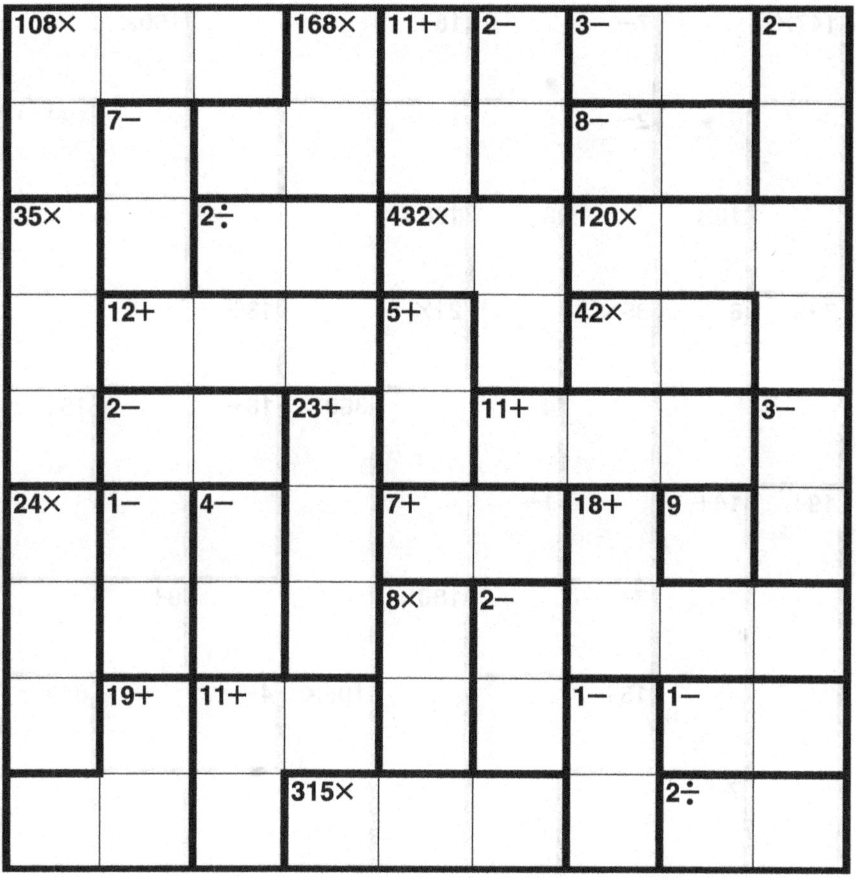

SOLVING
TIME

55

\+ − × ÷

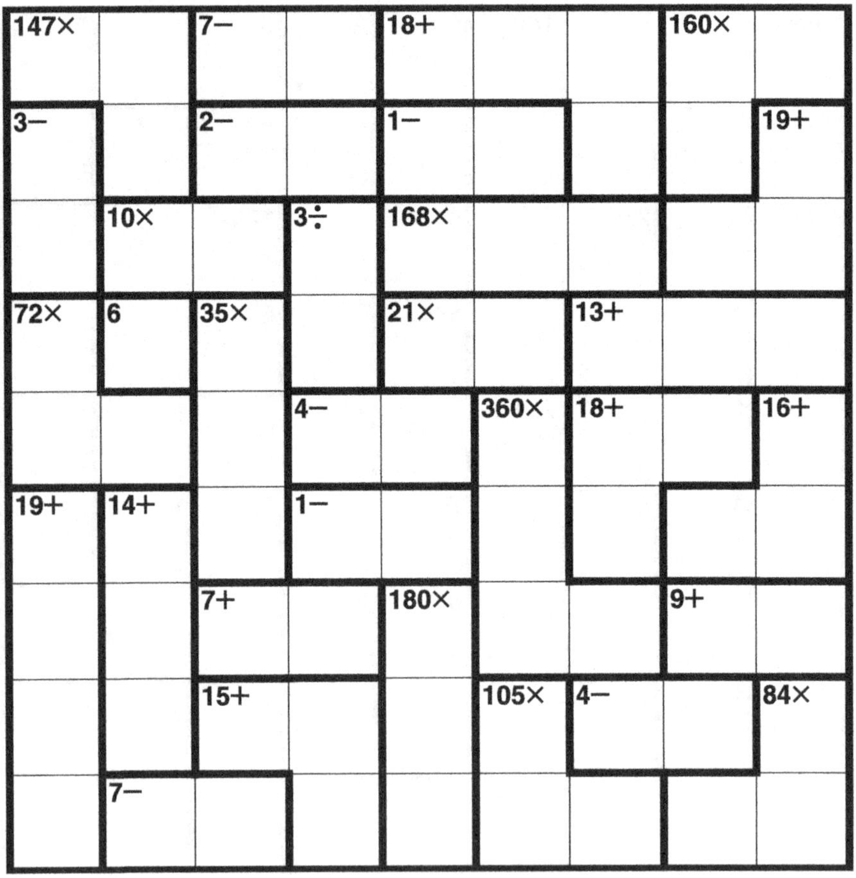

SOLVING
TIME

56

+ − × ÷

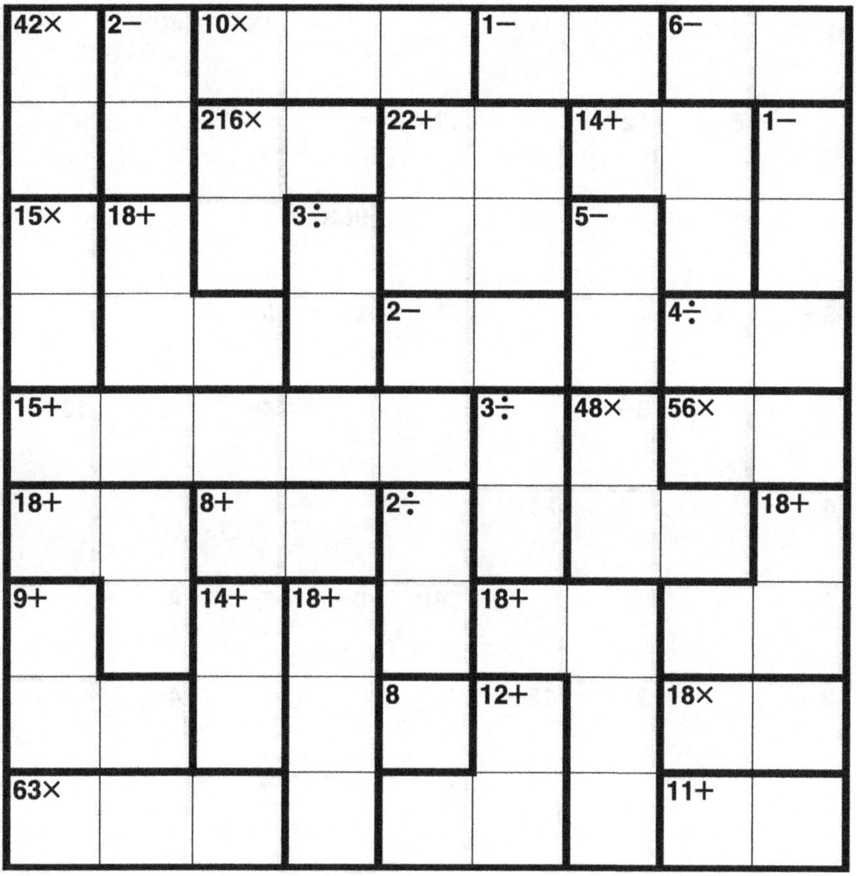

SOLVING

TIME

57

+ − × ÷

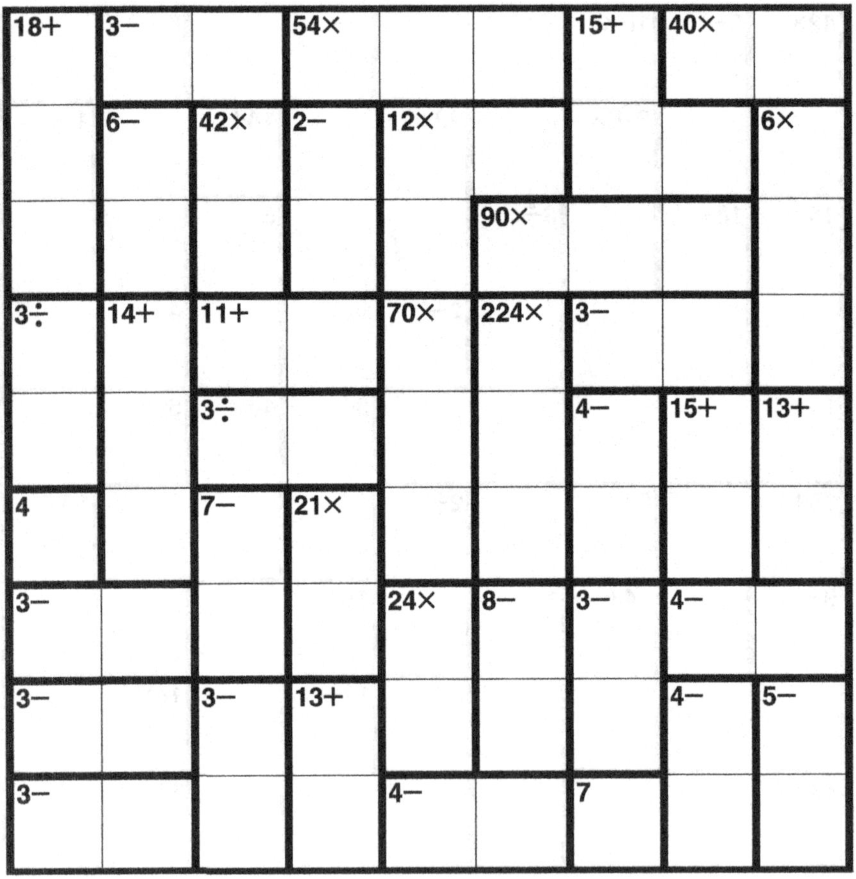

SOLVING TIME

58

+ − × ÷

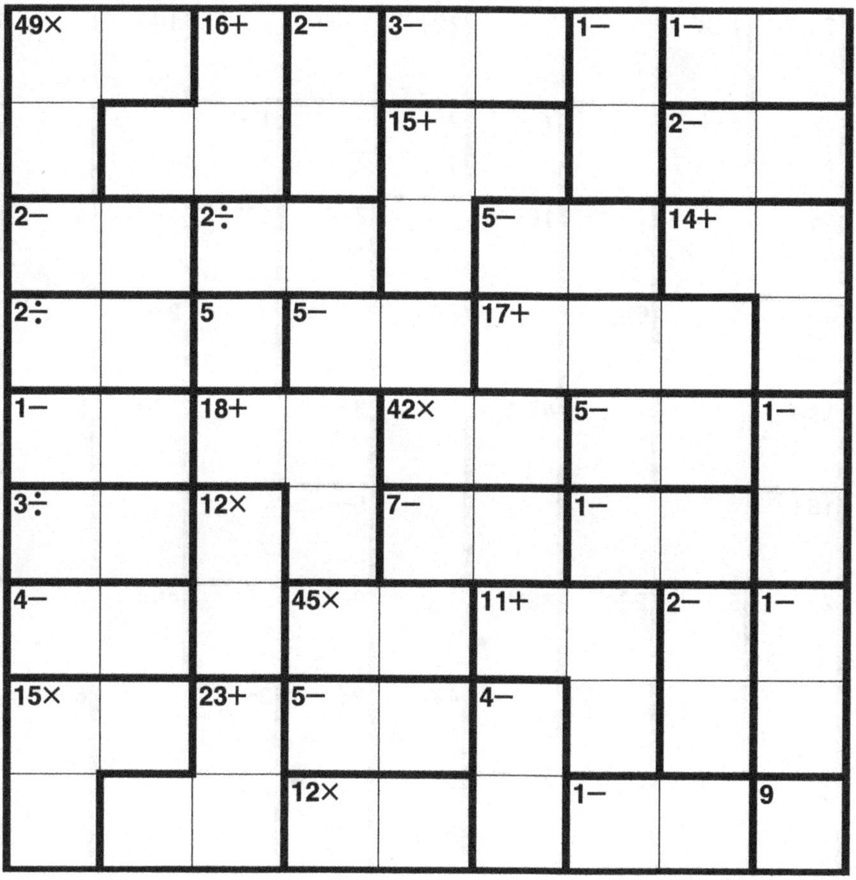

SOLVING
TIME

59

+ − × ÷

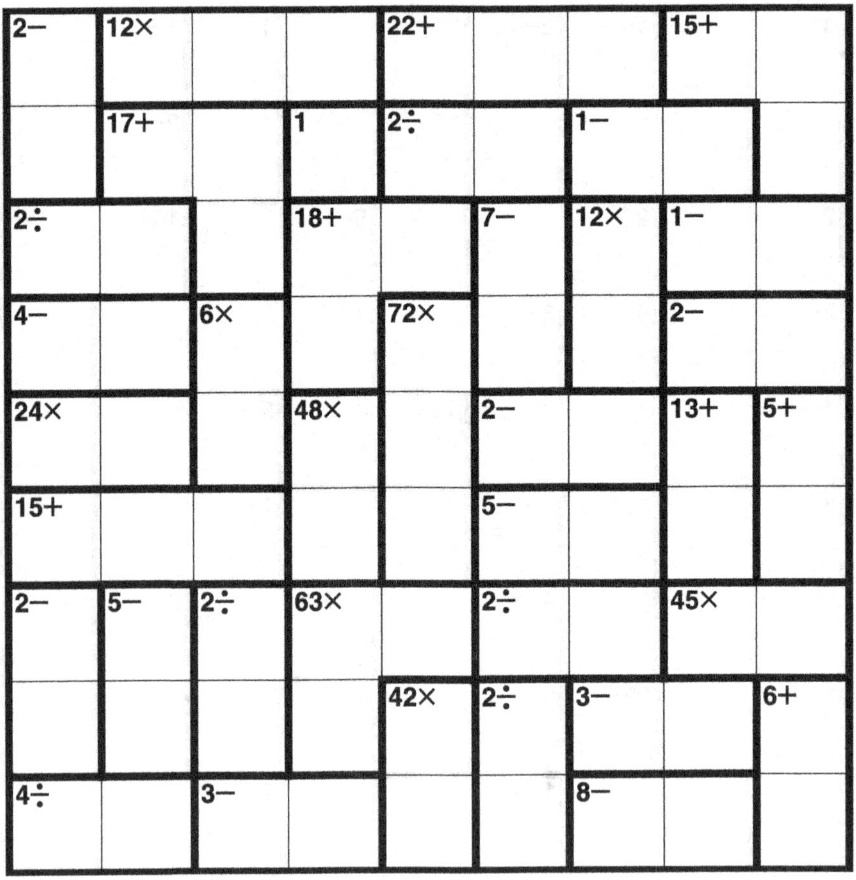

SOLVING
TIME

60

+ − × ÷

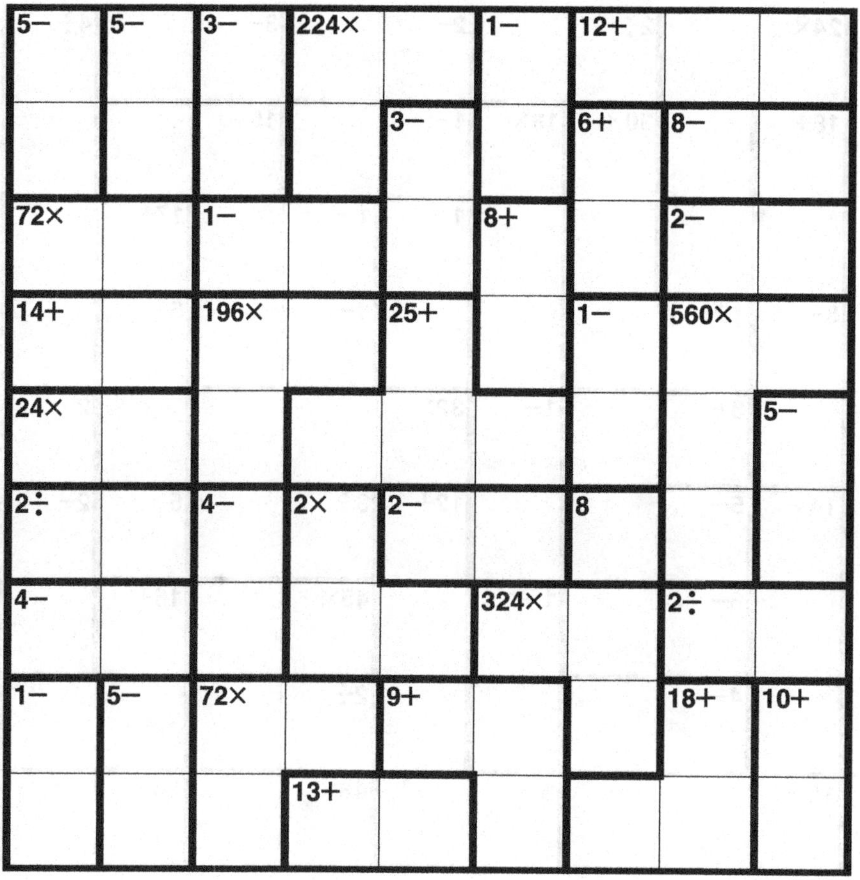

SOLVING
TIME

61

+ − × ÷

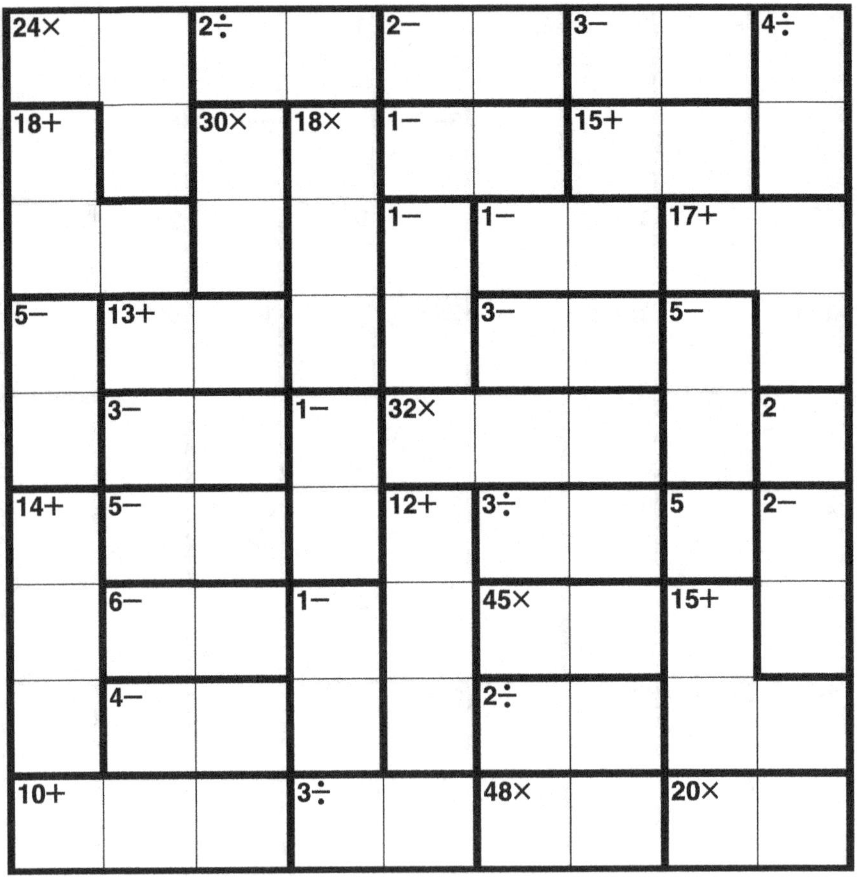

SOLVING
TIME

62

+ − × ÷

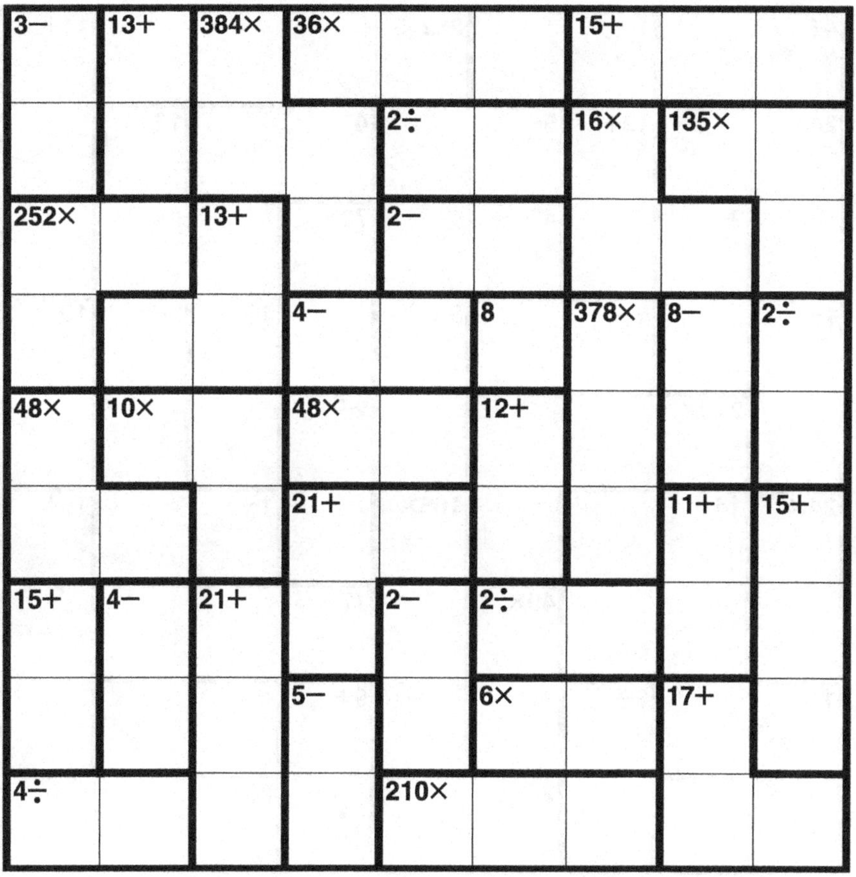

SOLVING
TIME

63

+ − × ÷

4÷		1−		252×				11+
24+		13+	5−		4−		12+	
	3−		4÷		72×			
1−		6−		3−		4÷	3−	10+
	240×				2÷			
24×	14+			105×		1−		16+
			40×		17+			
1−		3÷			9+	4−		
1−			2÷			288×		

SOLVING TIME

64

+ − × ÷

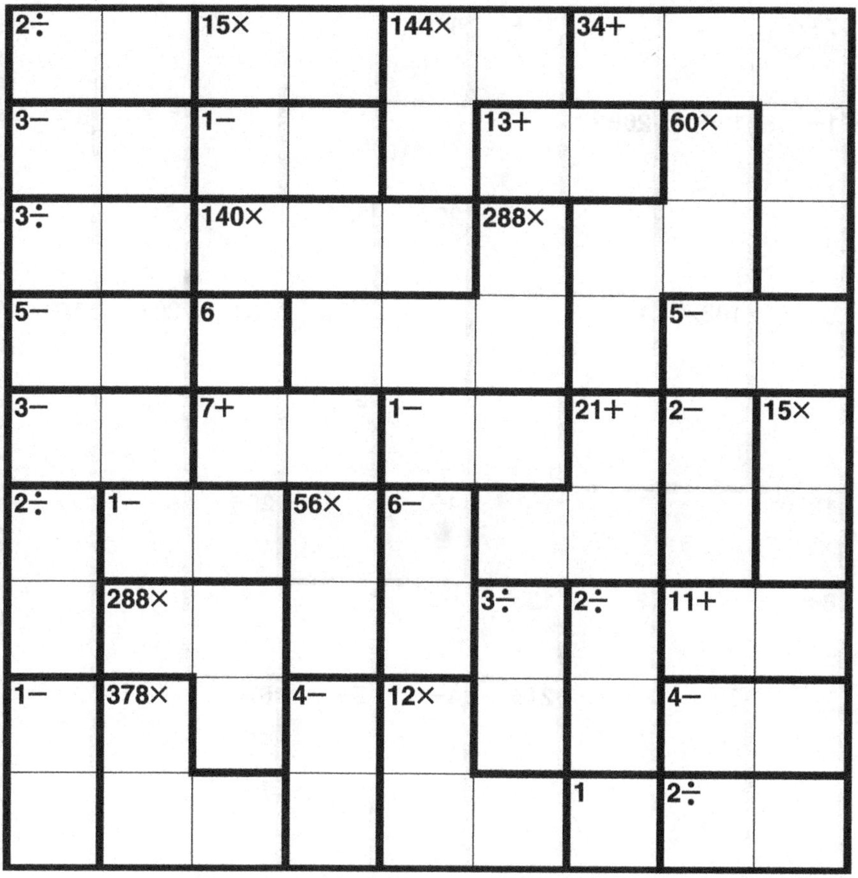

SOLVING

TIME

65

+ − × ÷

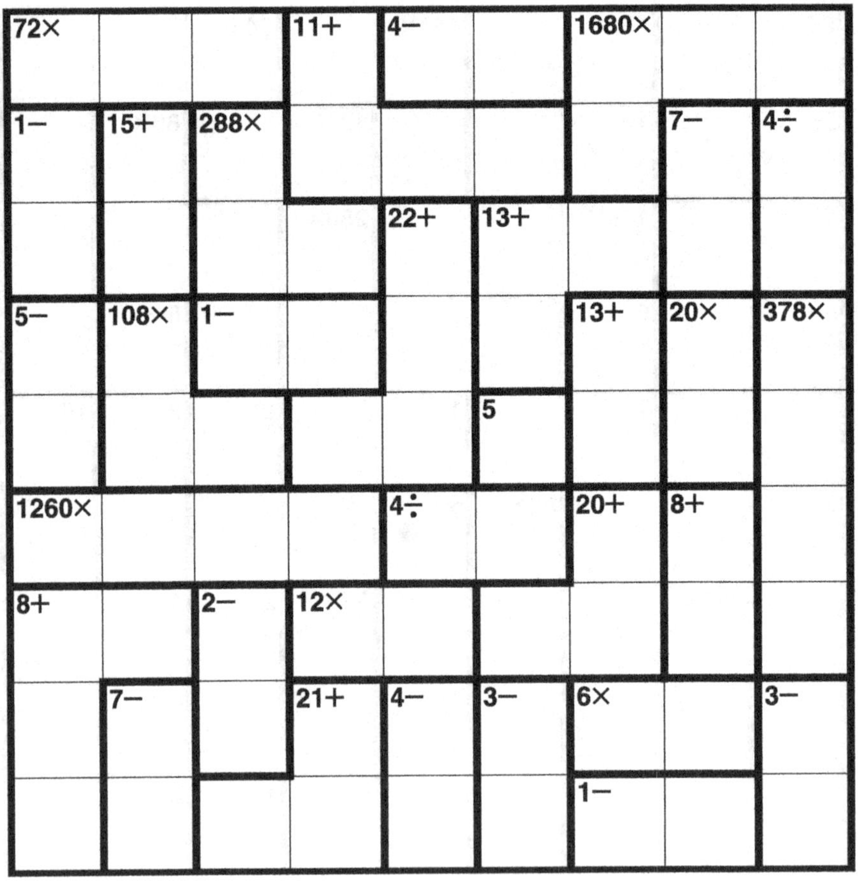

SOLVING
TIME

66

+ − × ÷

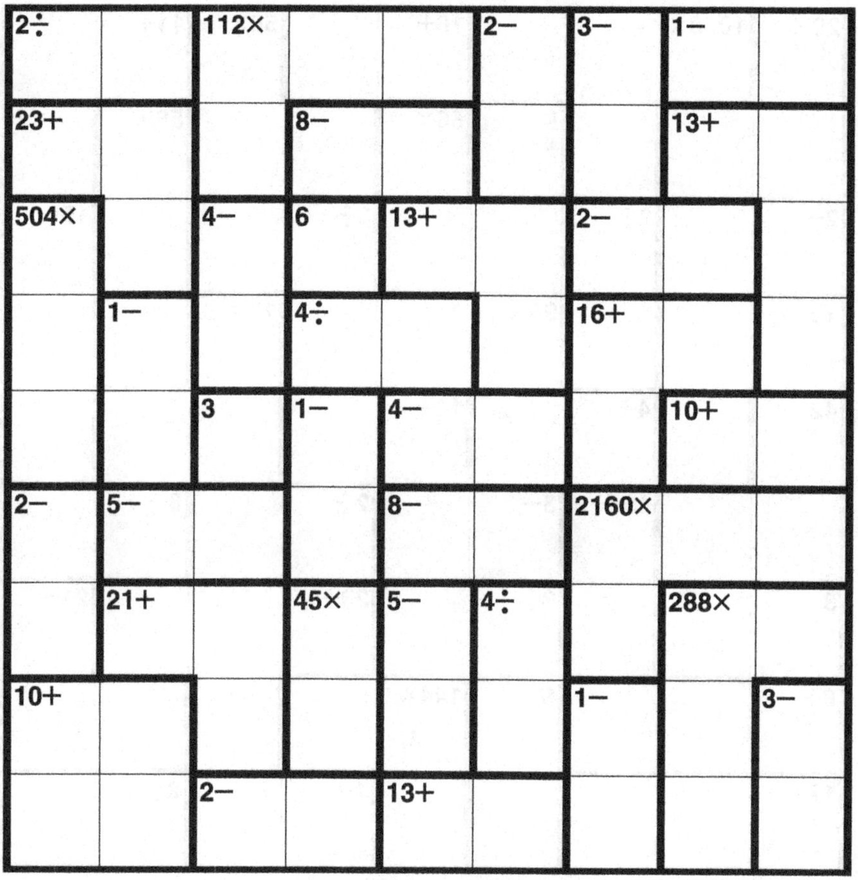

SOLVING
TIME

67

+ − × ÷

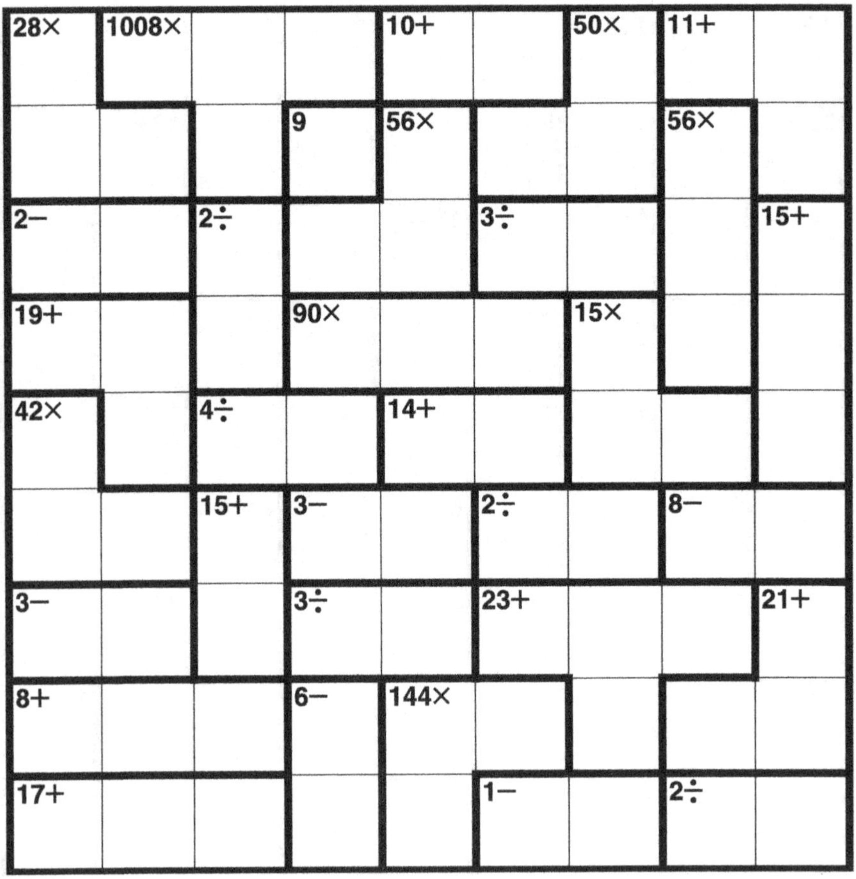

SOLVING
TIME

68

+ − × ÷

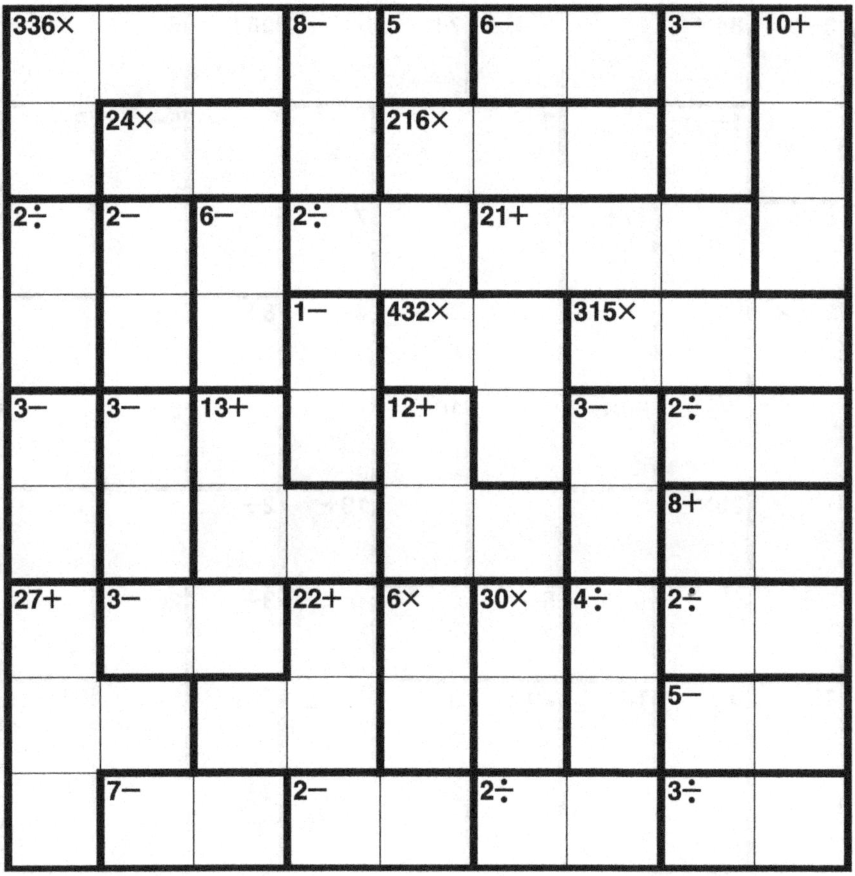

SOLVING
TIME

69

+ − × ÷

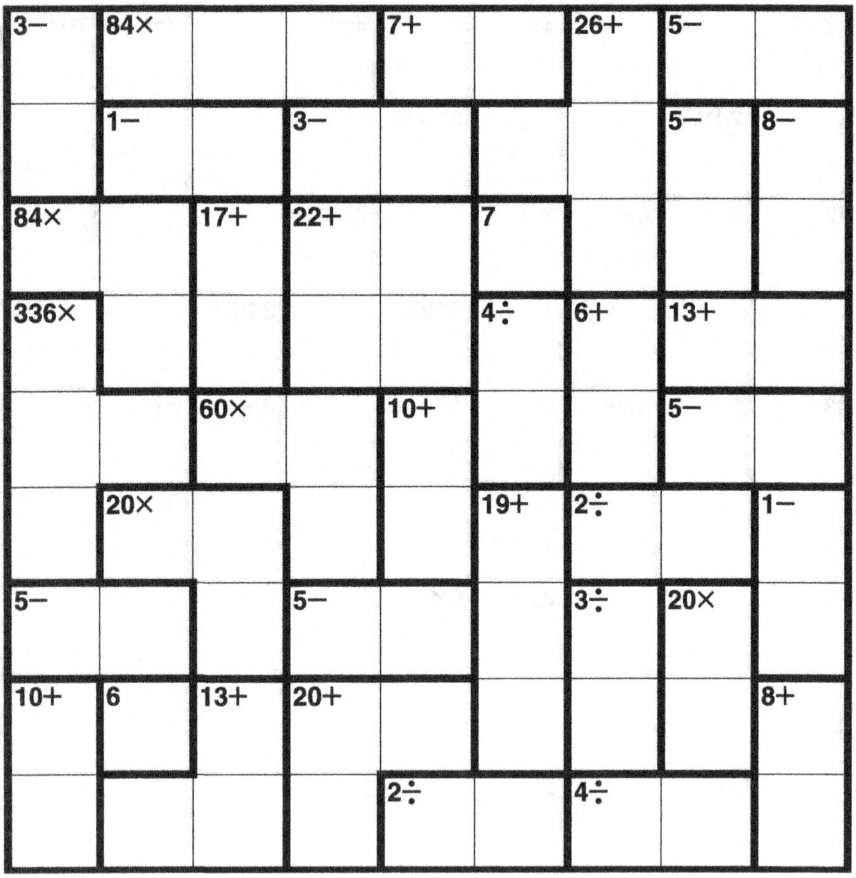

SOLVING TIME

70

+ − × ÷

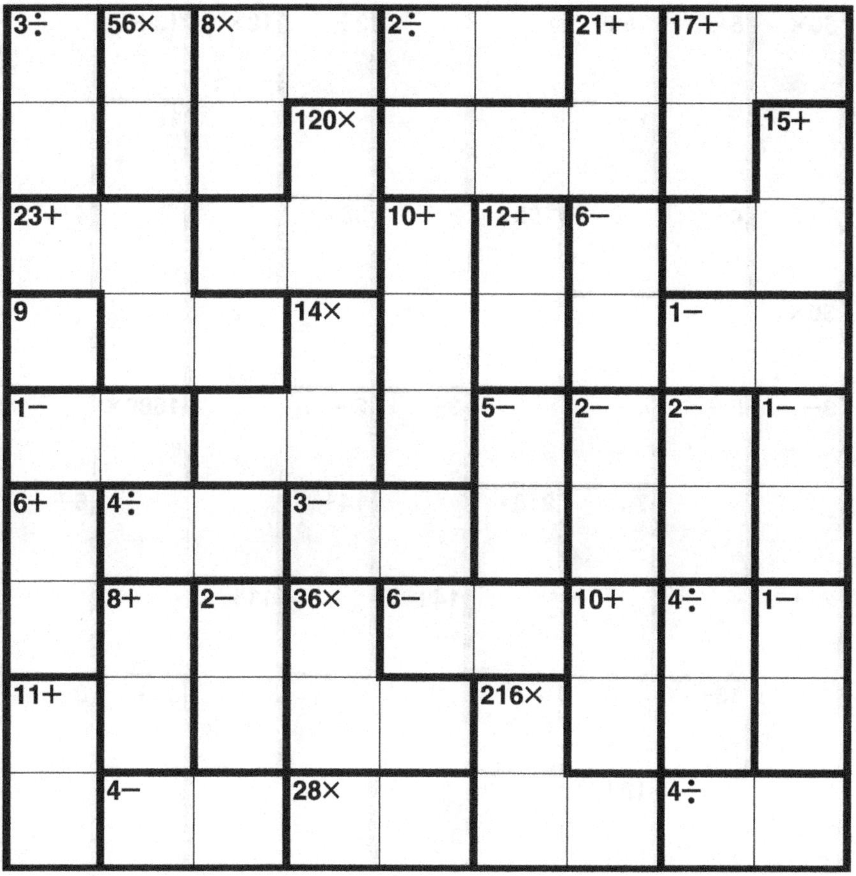

SOLVING
TIME

71

+ − × ÷

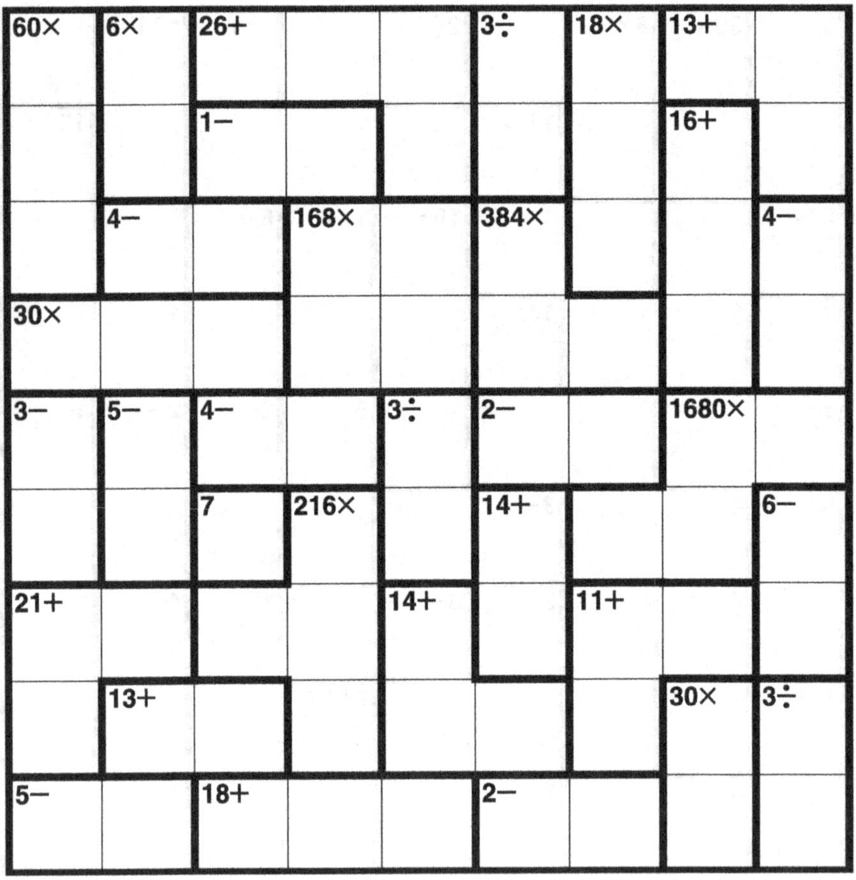

SOLVING
TIME

72

+ − × ÷

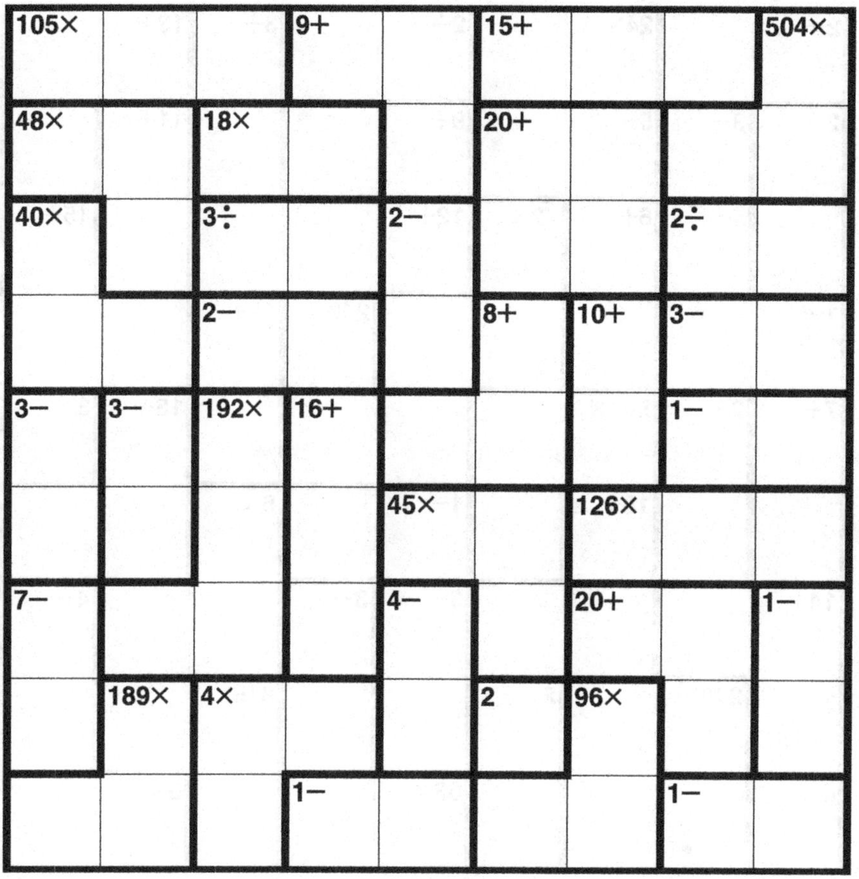

SOLVING
TIME

73

+ − × ÷

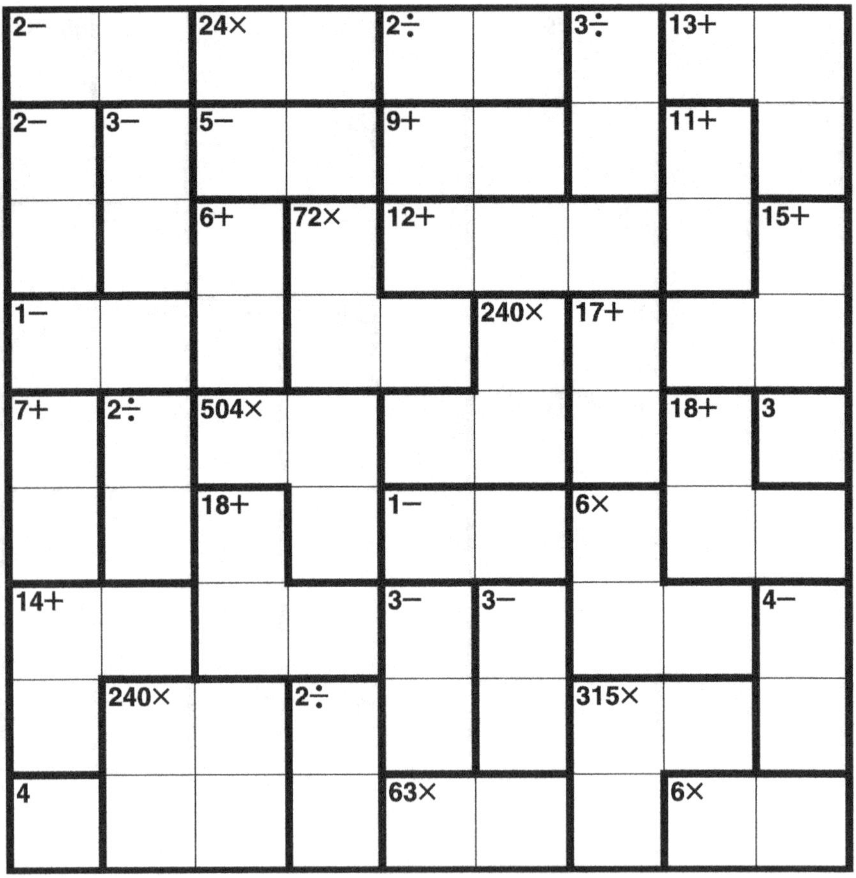

SOLVING
TIME

74

+ − × ÷

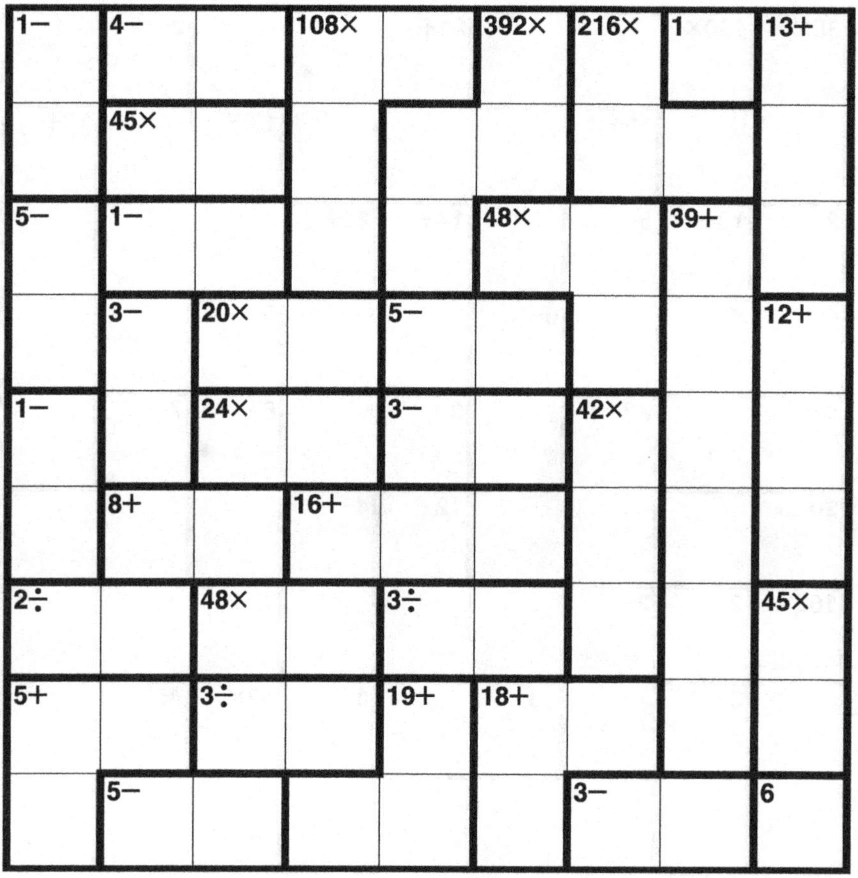

SOLVING
TIME

75

+ − × ÷

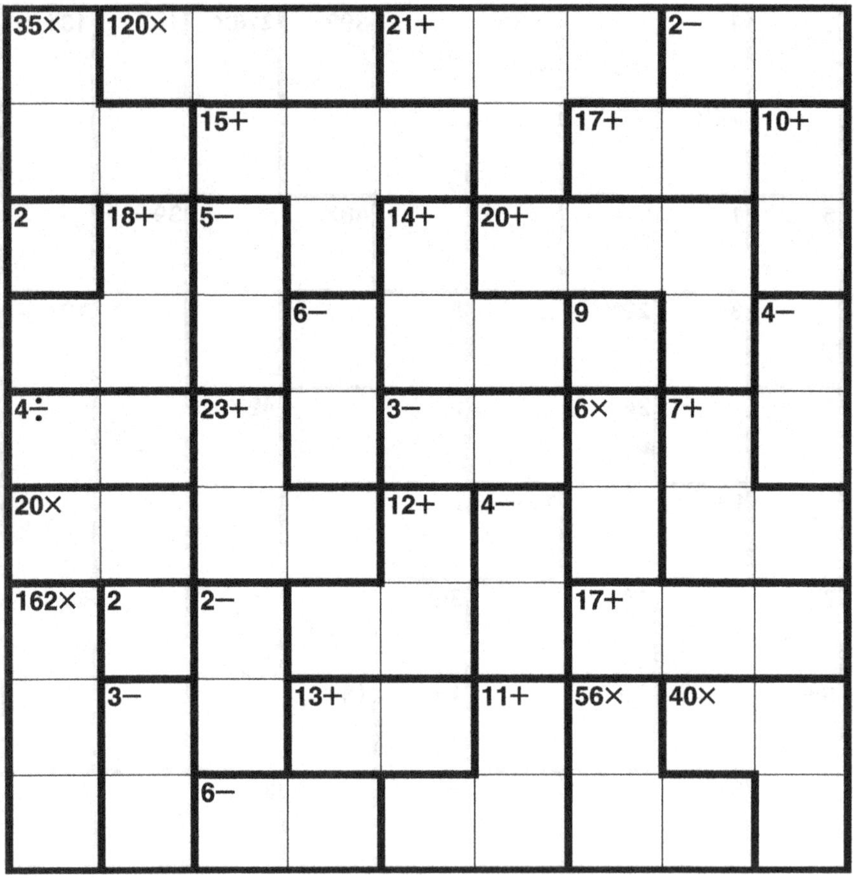

SOLVING TIME

76

+ − × ÷

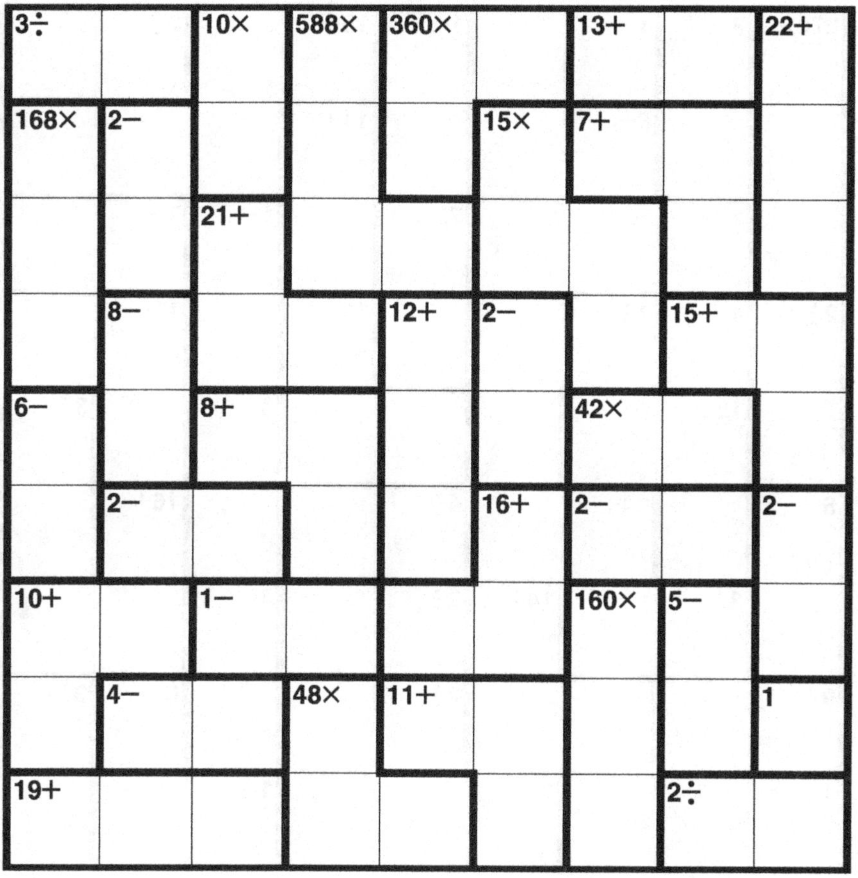

SOLVING
TIME

77

+ − × ÷

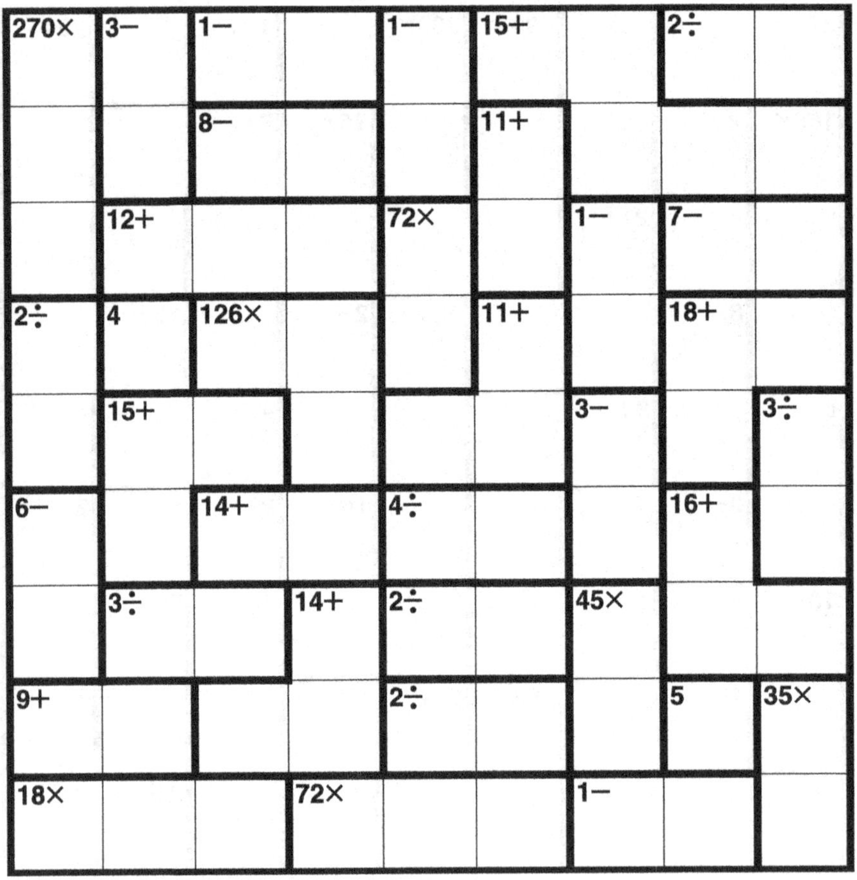

SOLVING TIME

78

+ − × ÷

9072×		7+		21+			24×	
			168×	7+				24+
2−	28×			10+		20+		
		270×			24+			
4÷	9+	4÷		7				
			2−	16+		135×	35×	
16×		2÷						1−
	180×		10+	3−				
7				14+			4÷	

SOLVING
TIME

79

+ − × ÷

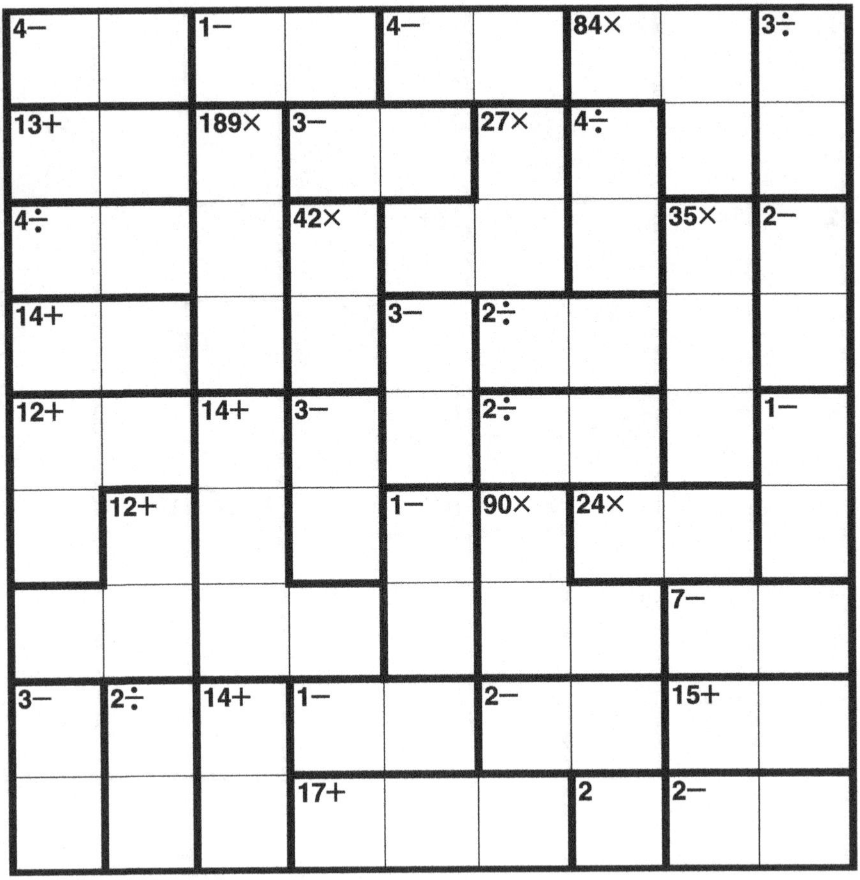

SOLVING TIME

80

+ − × ÷

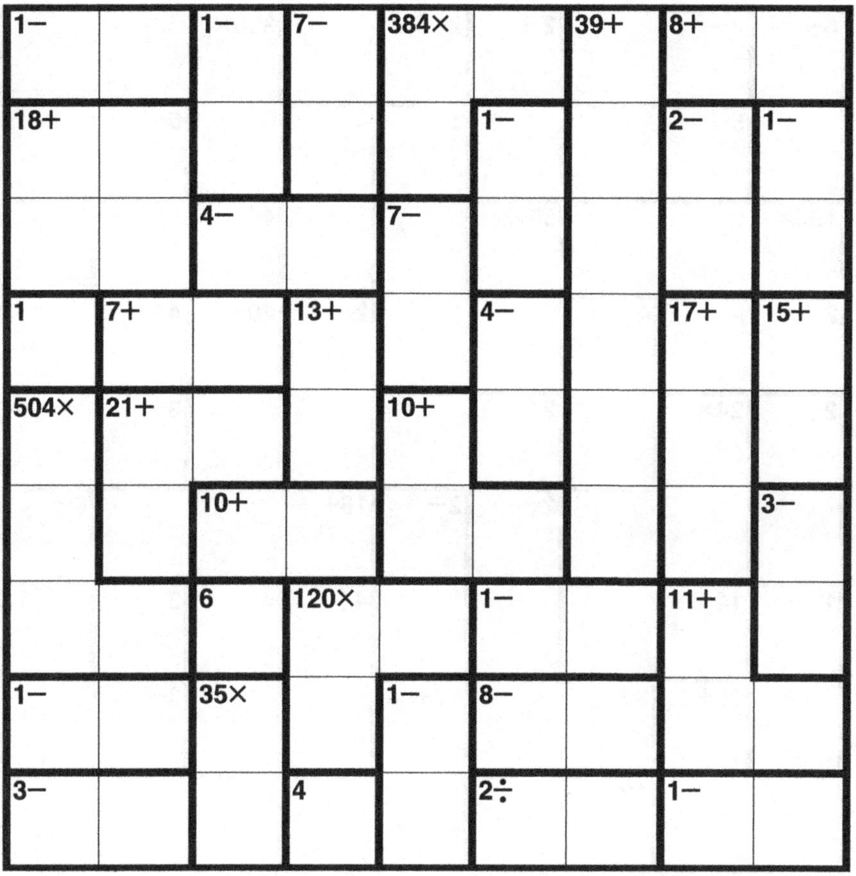

SOLVING
TIME

81

+ − × ÷

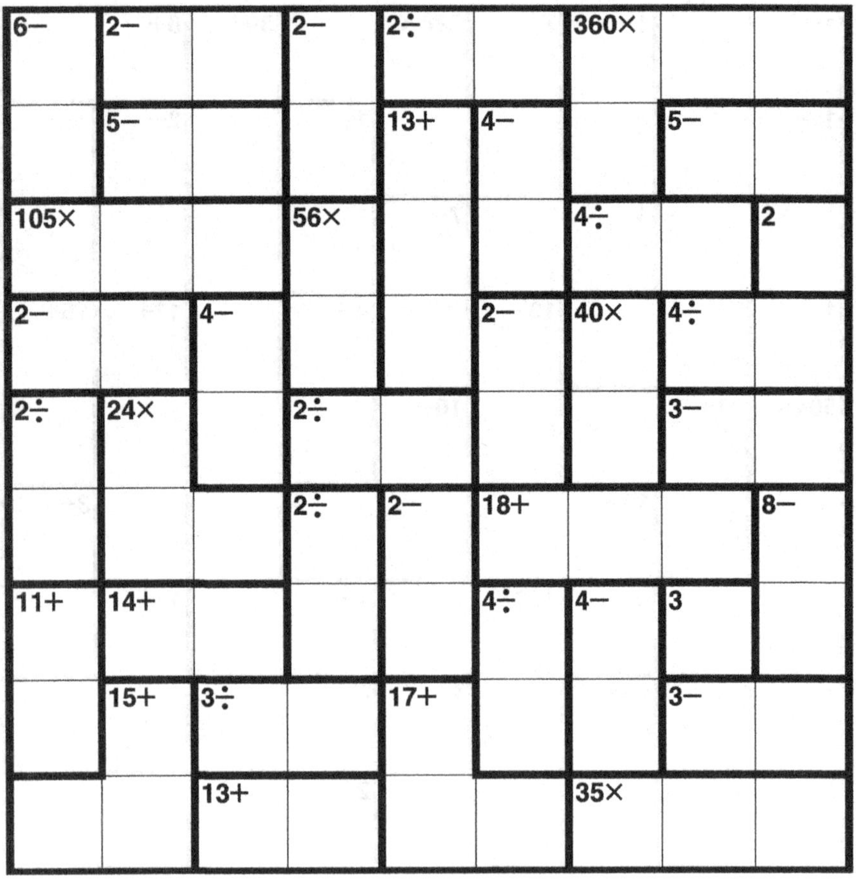

SOLVING
TIME

82

+ − × ÷

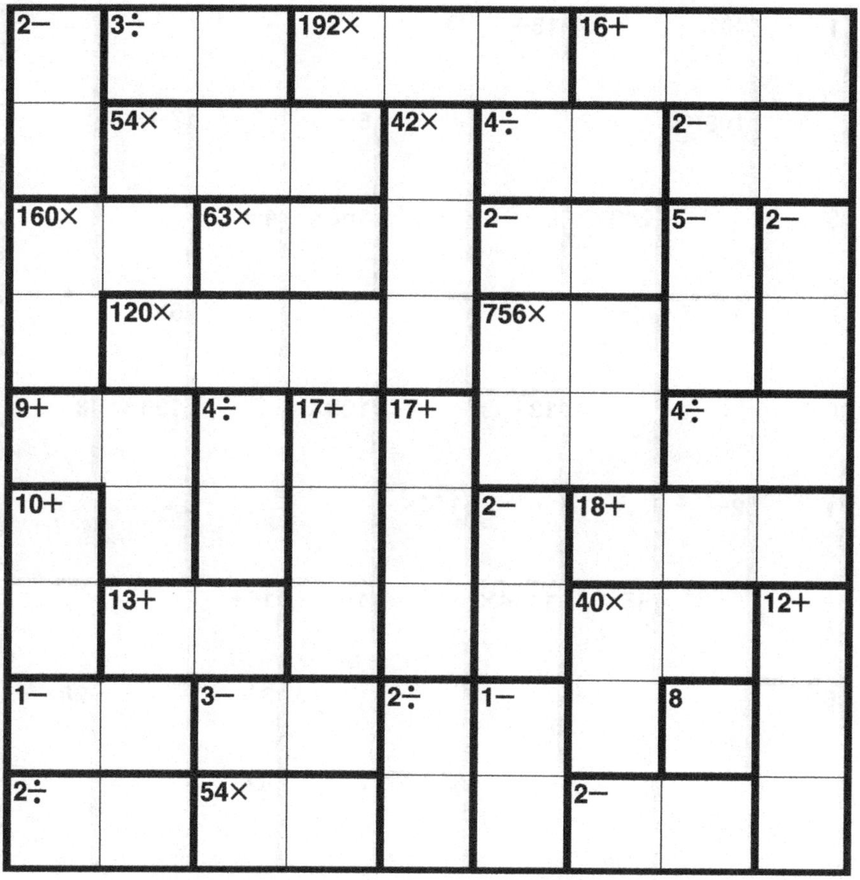

SOLVING TIME

83

+ − × ÷

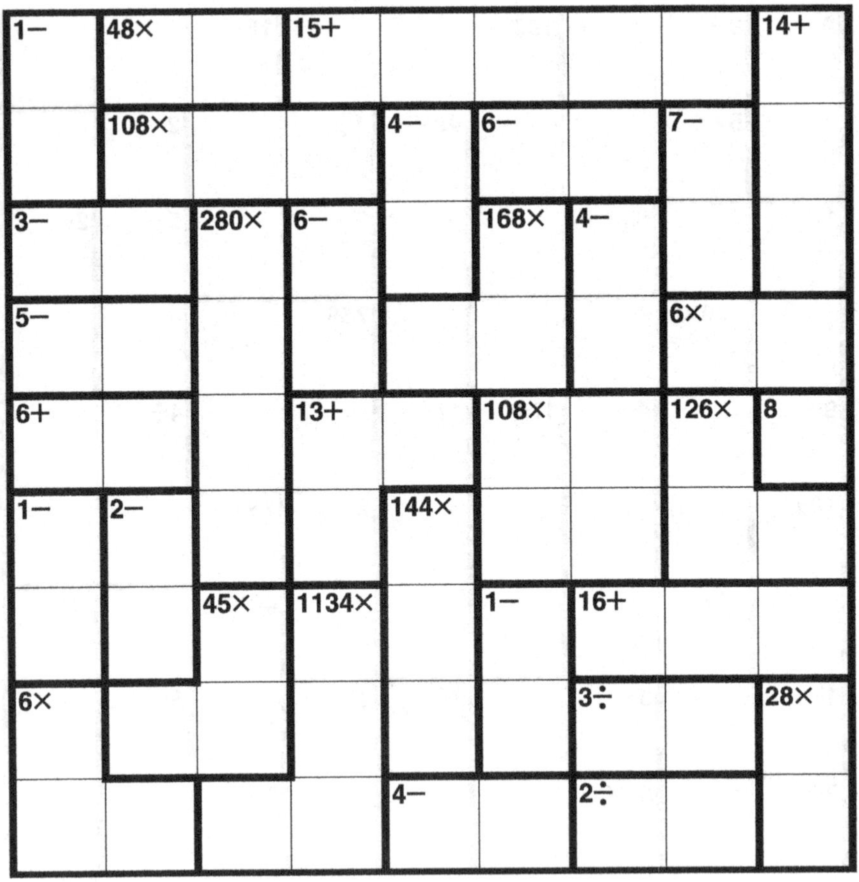

SOLVING
TIME

84

+ − × ÷

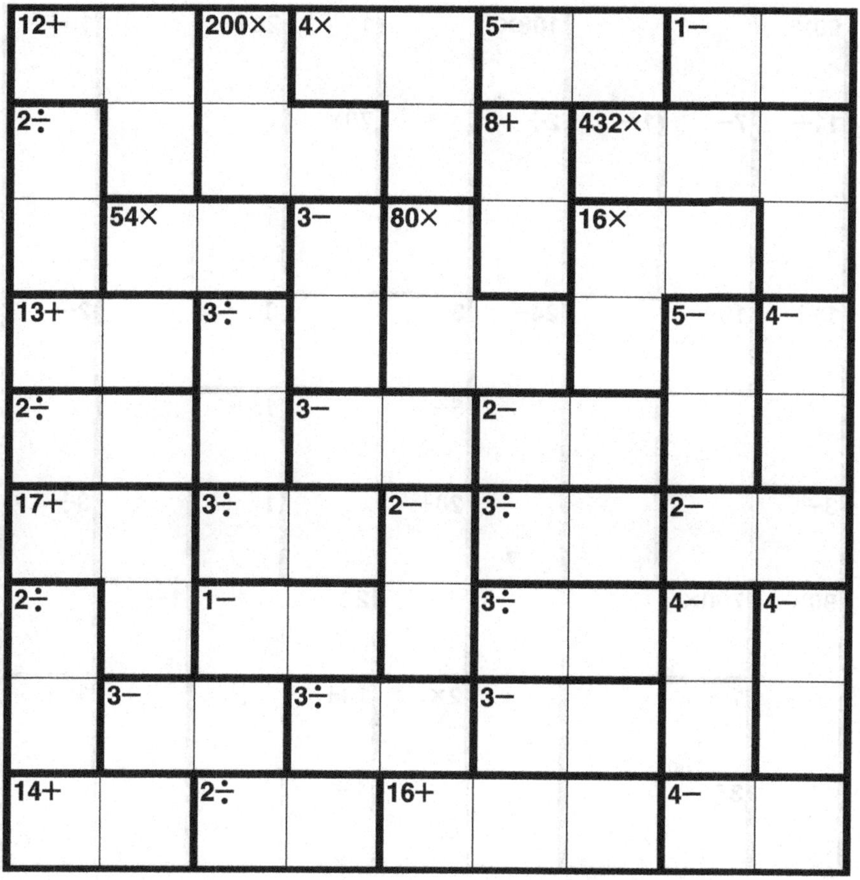

SOLVING
TIME

85

+ − × ÷

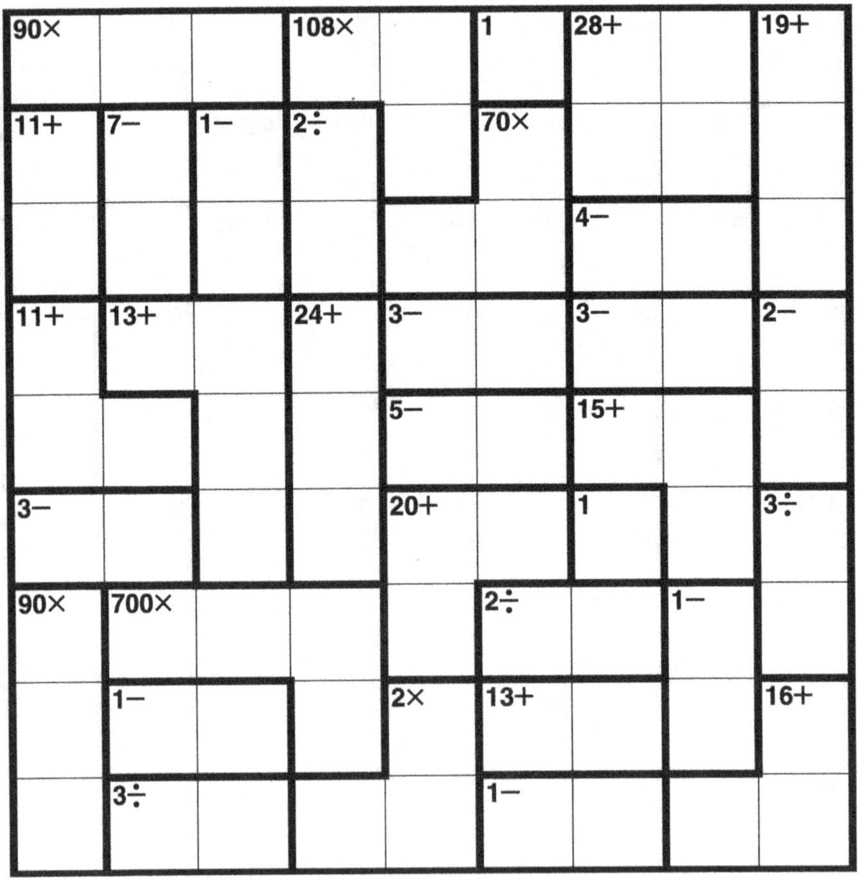

SOLVING
TIME

86

+ − × ÷

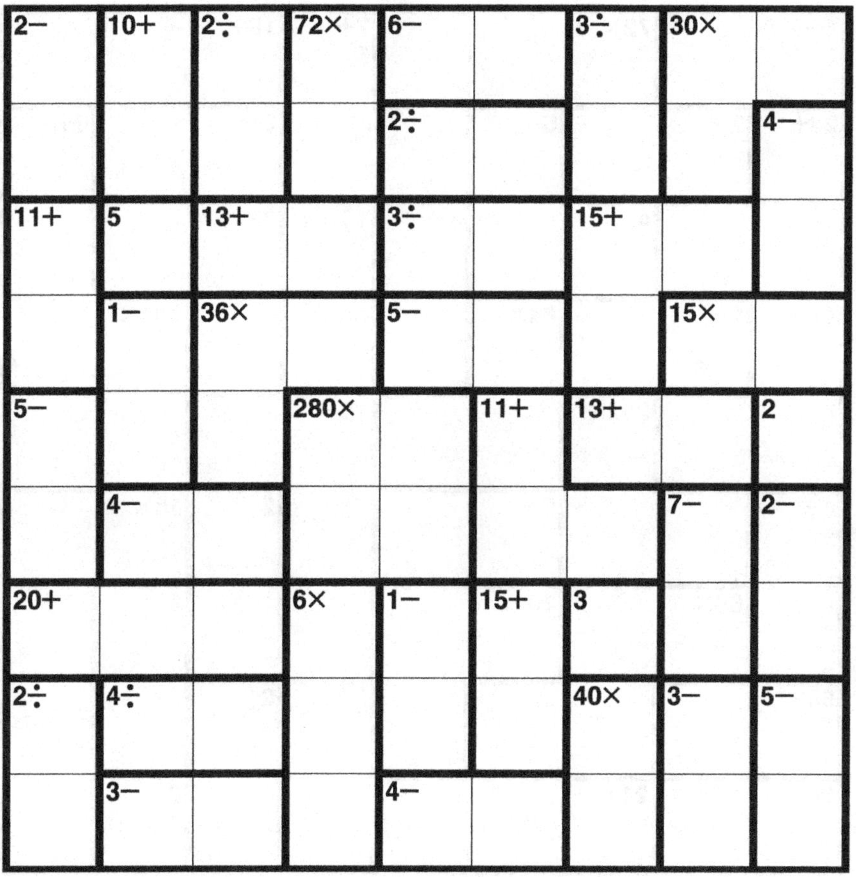

SOLVING
TIME

87

+ − × ÷

1−		72×			7+	18+		
23+	2−		6−			3−		9+
		4		3−	15+	3×		
11+	1−		84×				384×	
		252×			15+			
4÷			5−			2−	15×	
	64×		13+				3−	16+
54×					3÷	2÷		
3−		21×					2	

SOLVING TIME

88

+ − × ÷

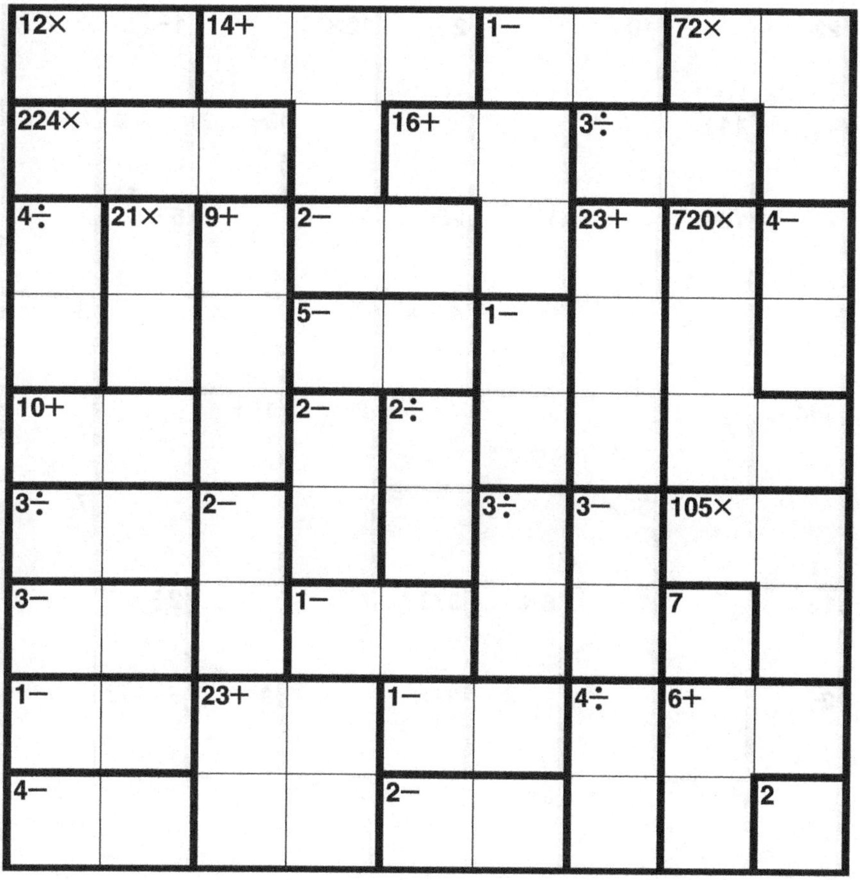

SOLVING
TIME

89

+ − × ÷

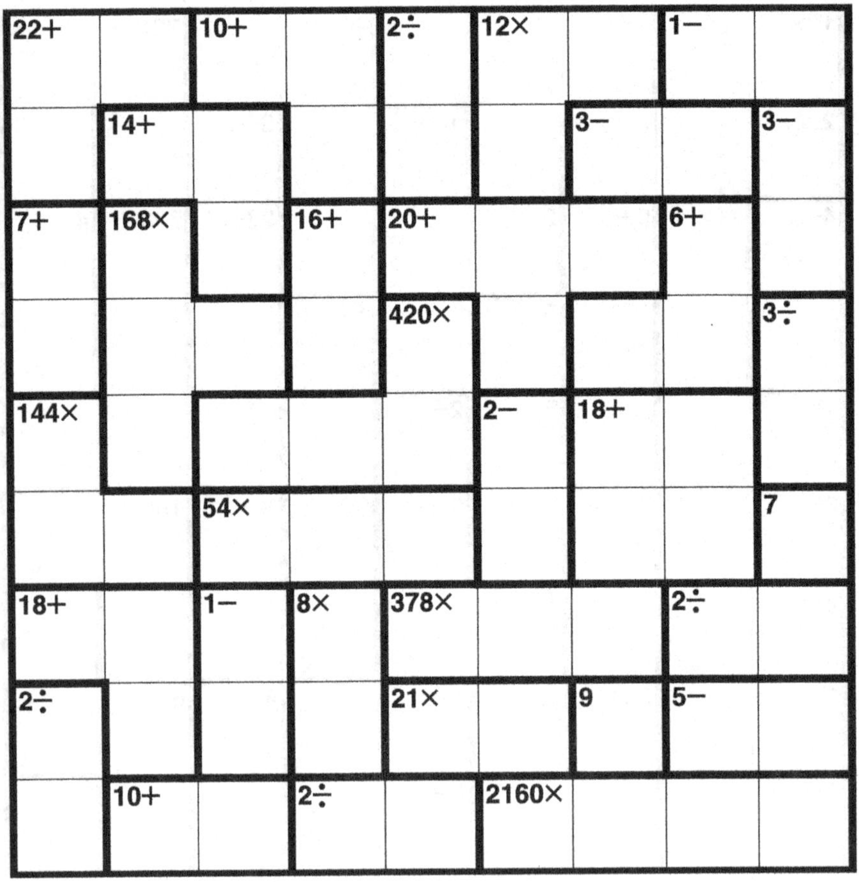

SOLVING
TIME

90

+ − × ÷

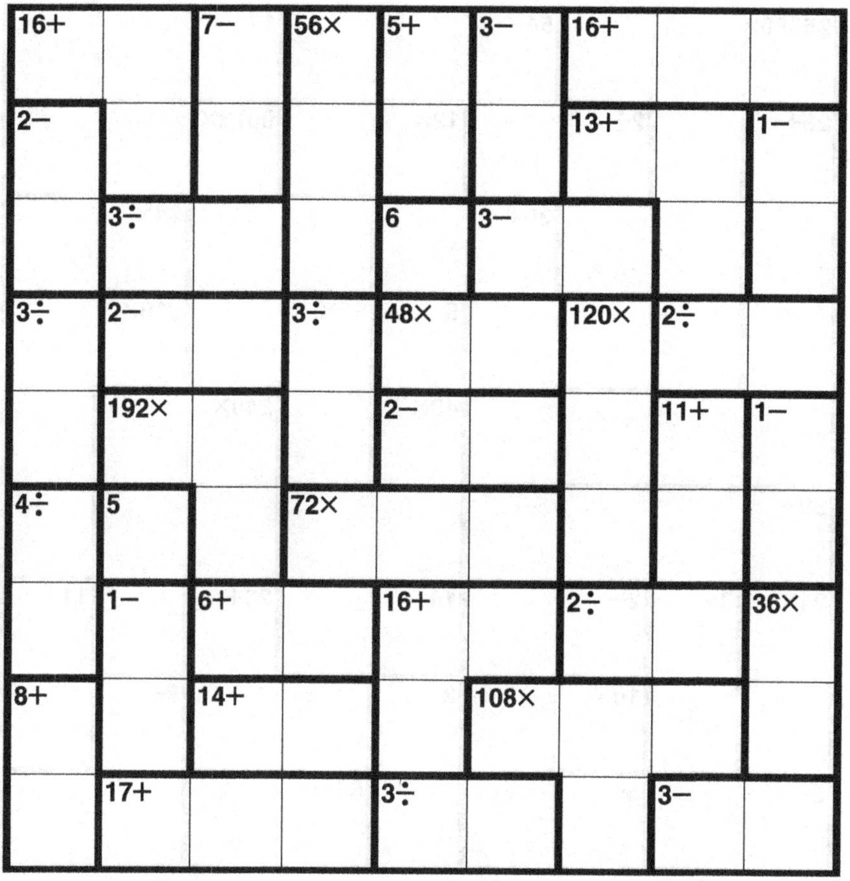

SOLVING
TIME

91

+ − × ÷

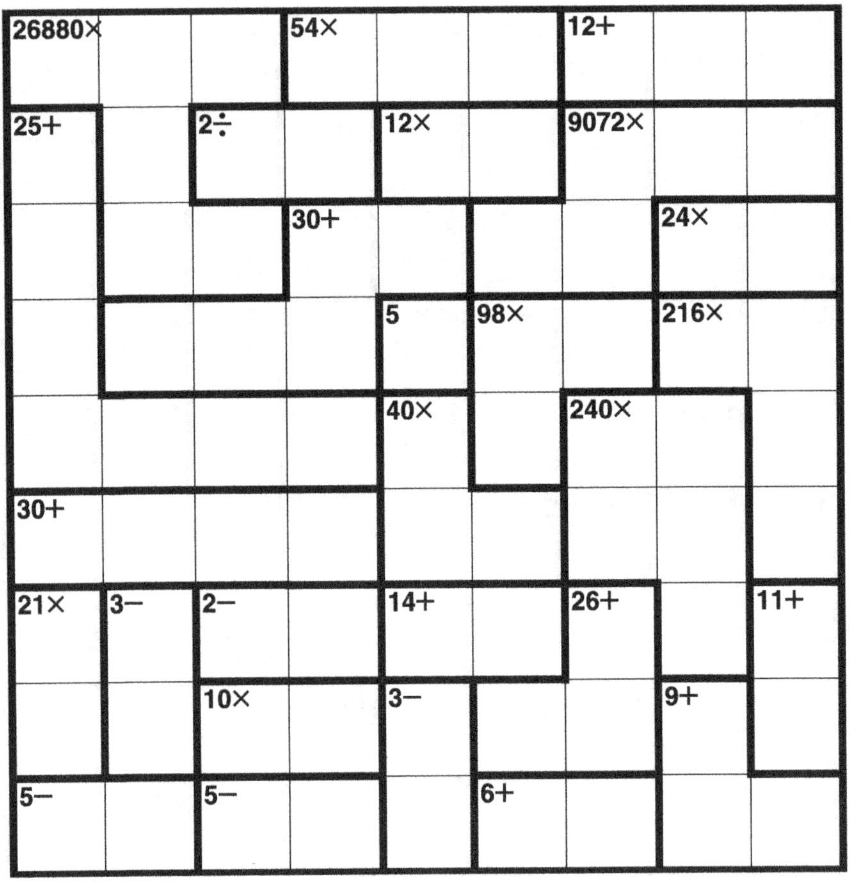

SOLVING
TIME

92

+ − × ÷

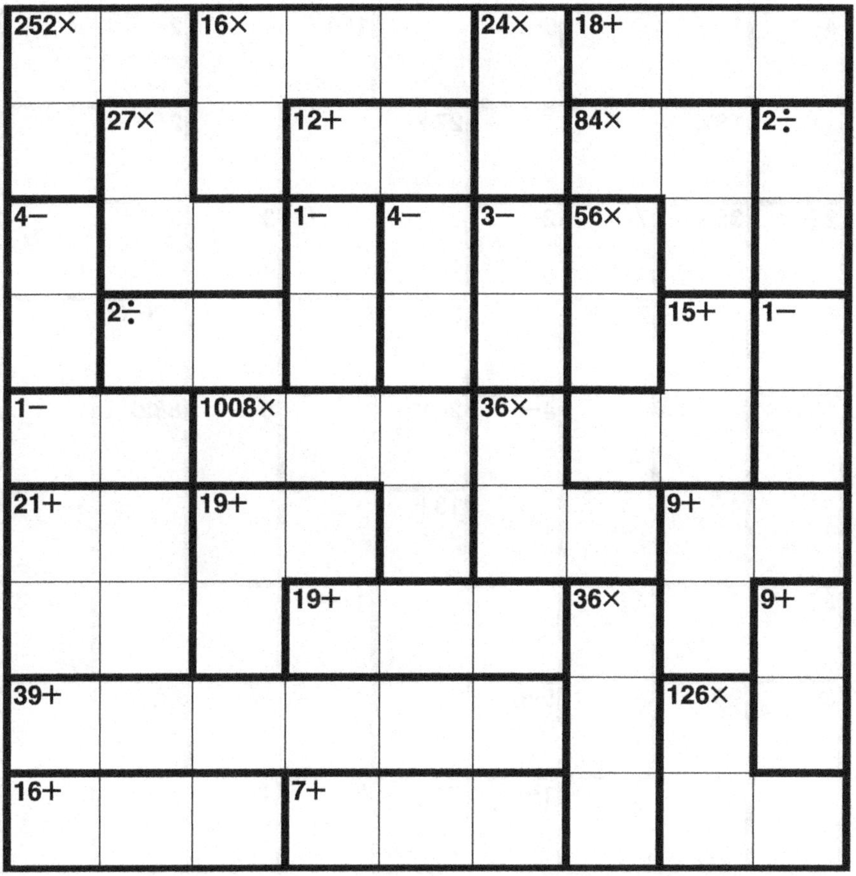

SOLVING
TIME

93

+ − × ÷

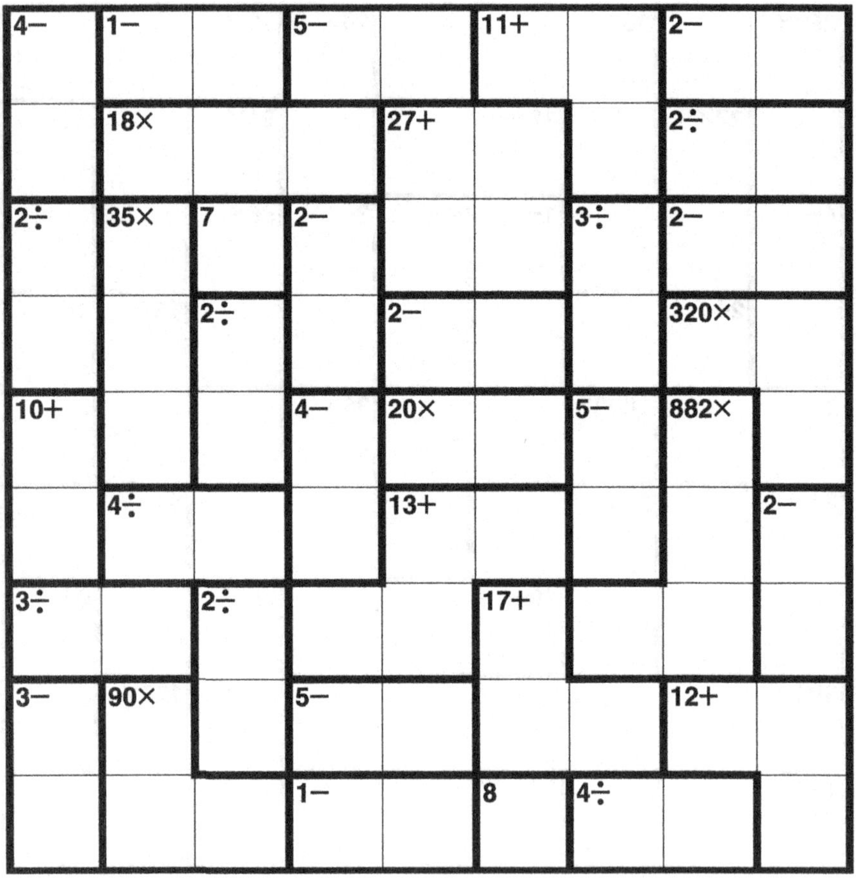

SOLVING
TIME

94

+ − × ÷

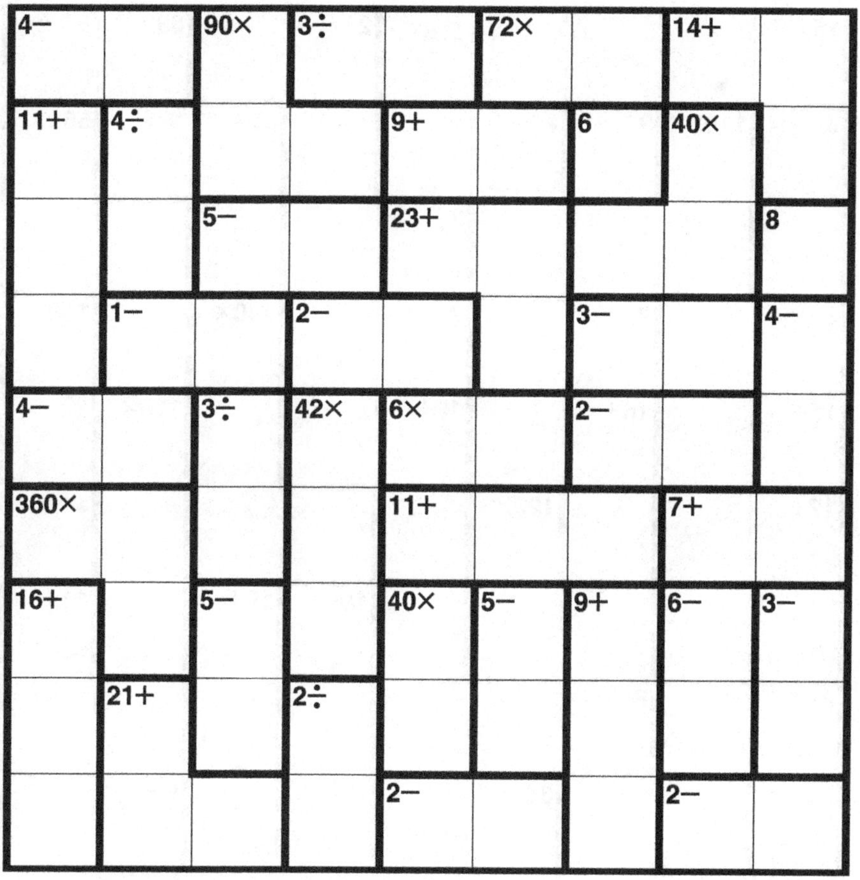

SOLVING TIME

95

+ − × ÷

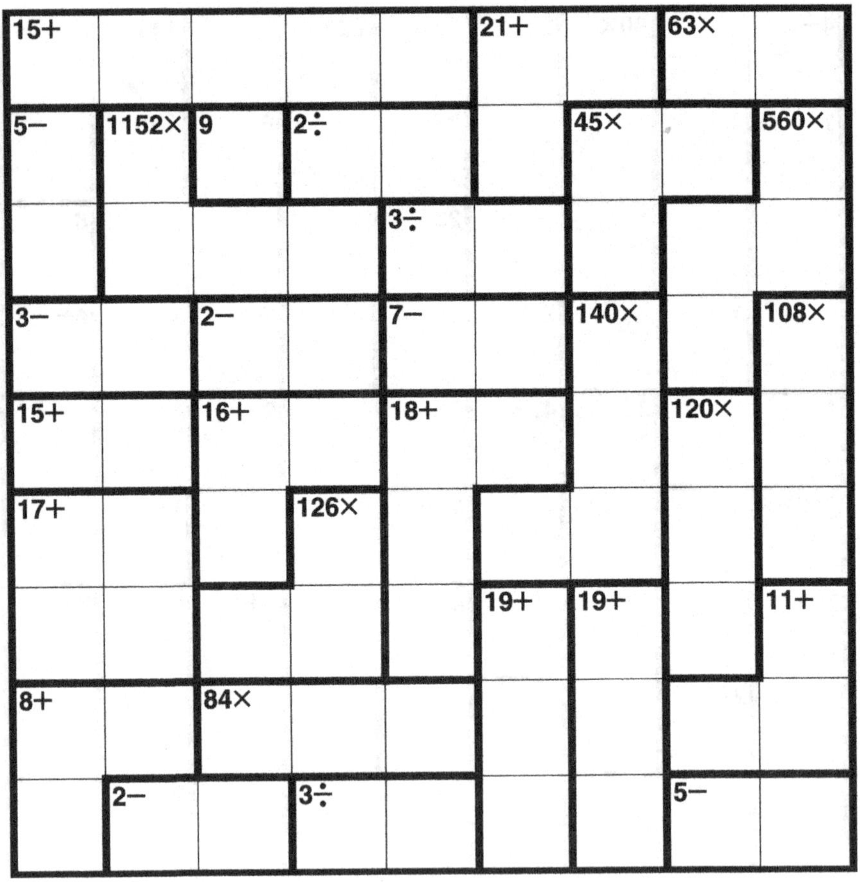

SOLVING TIME

96

$+ - \times \div$

2−		96×	5−		18+		12+	3−
5−				192×		6−		
15+	245×		6					
	20×		16×	15+		17+		
4÷							1008×	
	10+		4−		15+		4−	
144×			2−					
	144×	315×		3÷	168×		2÷	
						10×		

SOLVING
TIME

97

+ − × ÷

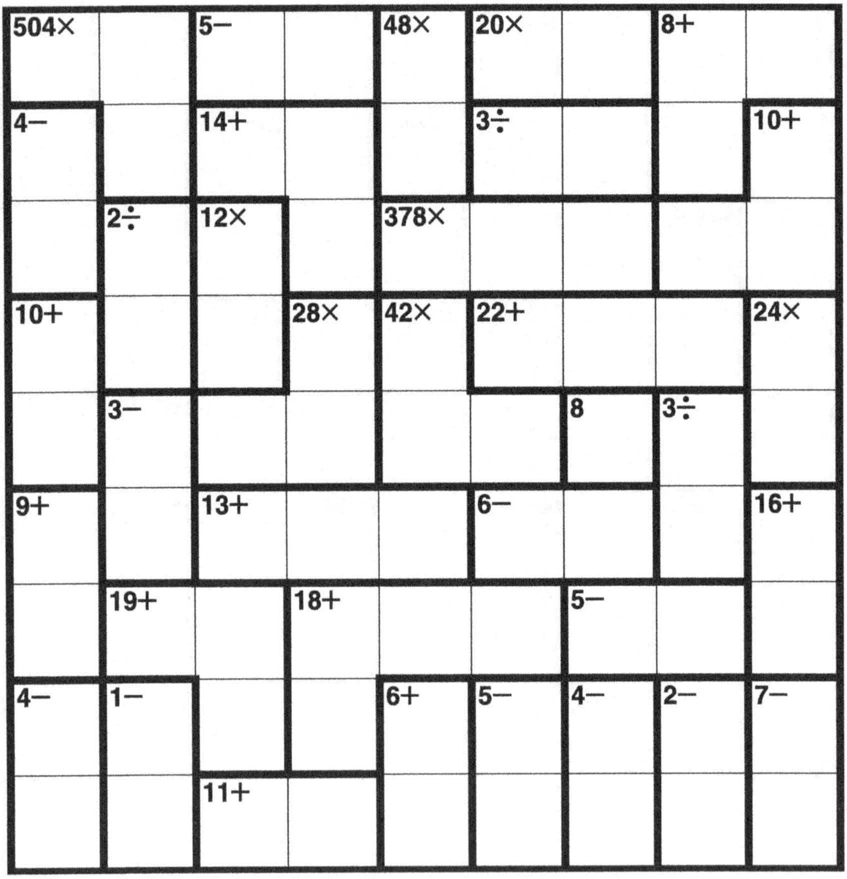

SOLVING
TIME

98

+ − × ÷

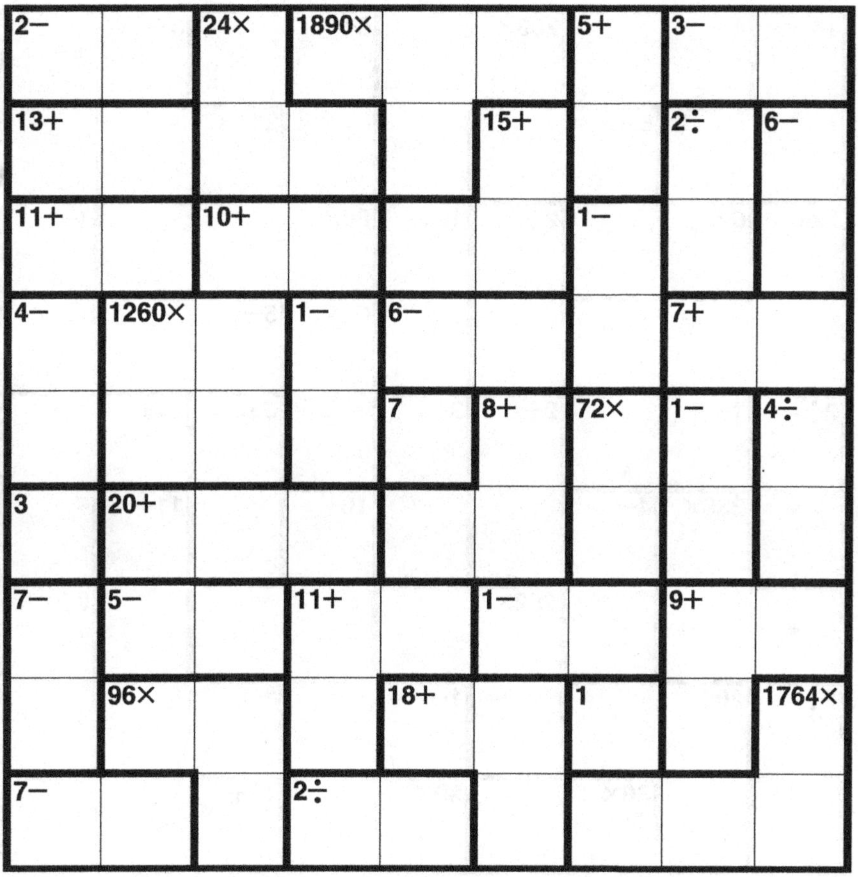

SOLVING TIME

99

+ − × ÷

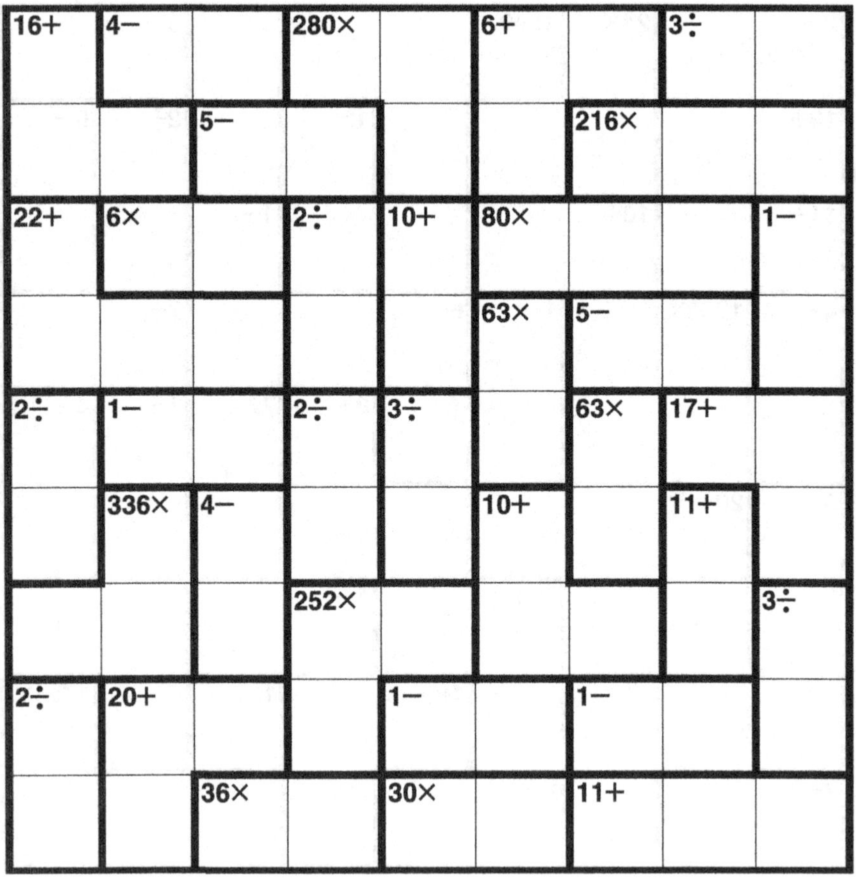

SOLVING
TIME

100

+ − × ÷

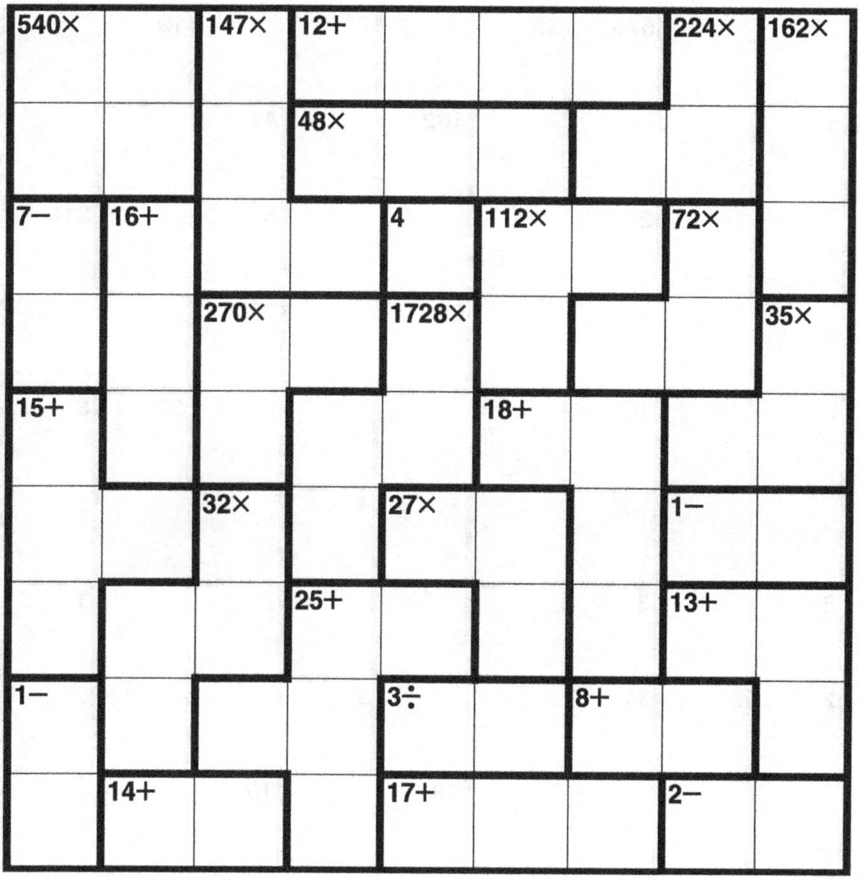

SOLVING
TIME

Bonus No-Op Puzzle! This puzzle does not include operations after the target numbers... you must deduce them! Note that the operation for cages with 3 or more squares must be addition or multiplication.

101

+ − × ÷

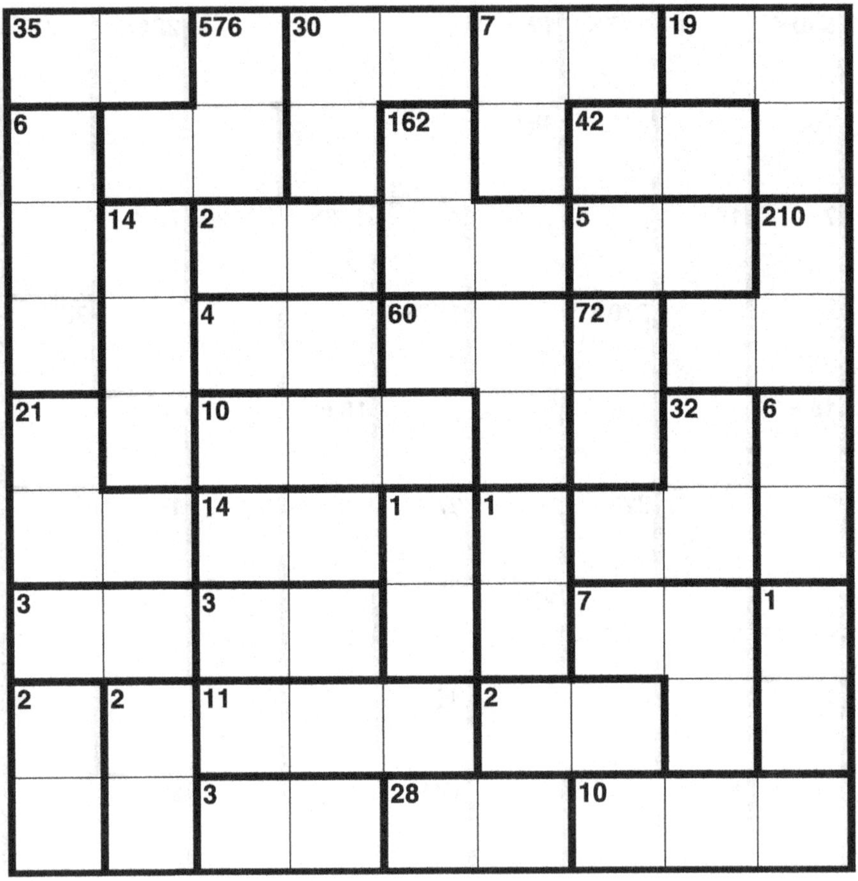

SOLVING
TIME

Bonus No-Op Puzzle! This puzzle does not include operations after the target numbers... you must deduce them! Note that the operation for cages with 3 or more squares must be addition or multiplication.

102

+ − × ÷

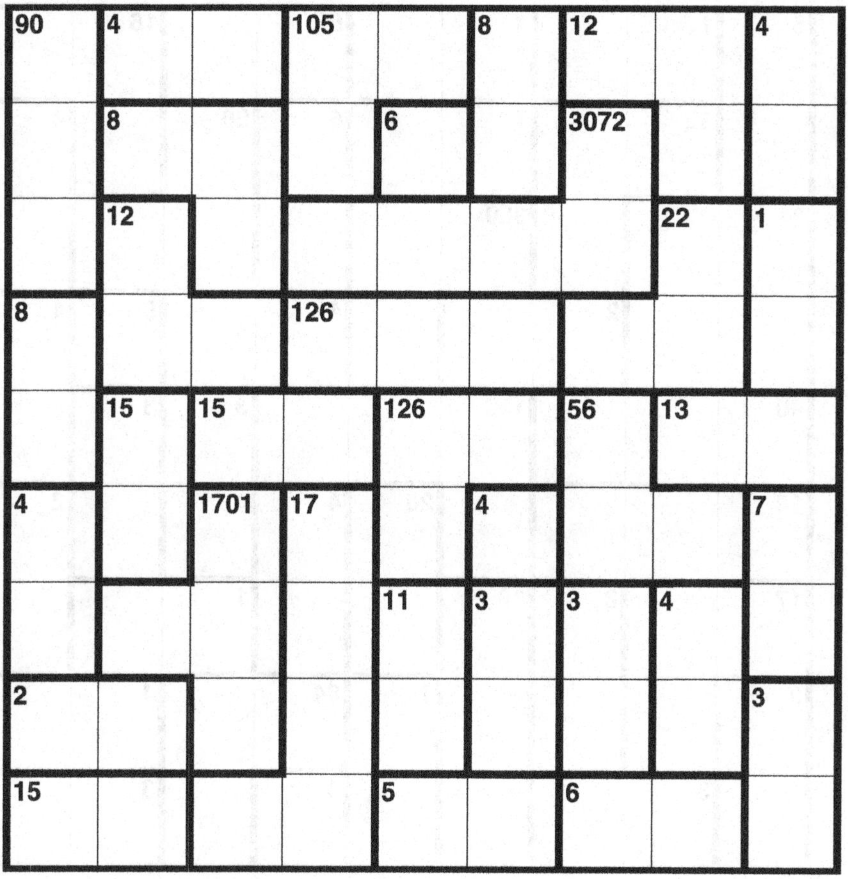

SOLVING

TIME

Bonus No-Op Puzzle! This puzzle does not include operations after the target numbers... you must deduce them! Note that the operation for cages with 3 or more squares must be addition or multiplication.

103

+ − × ÷

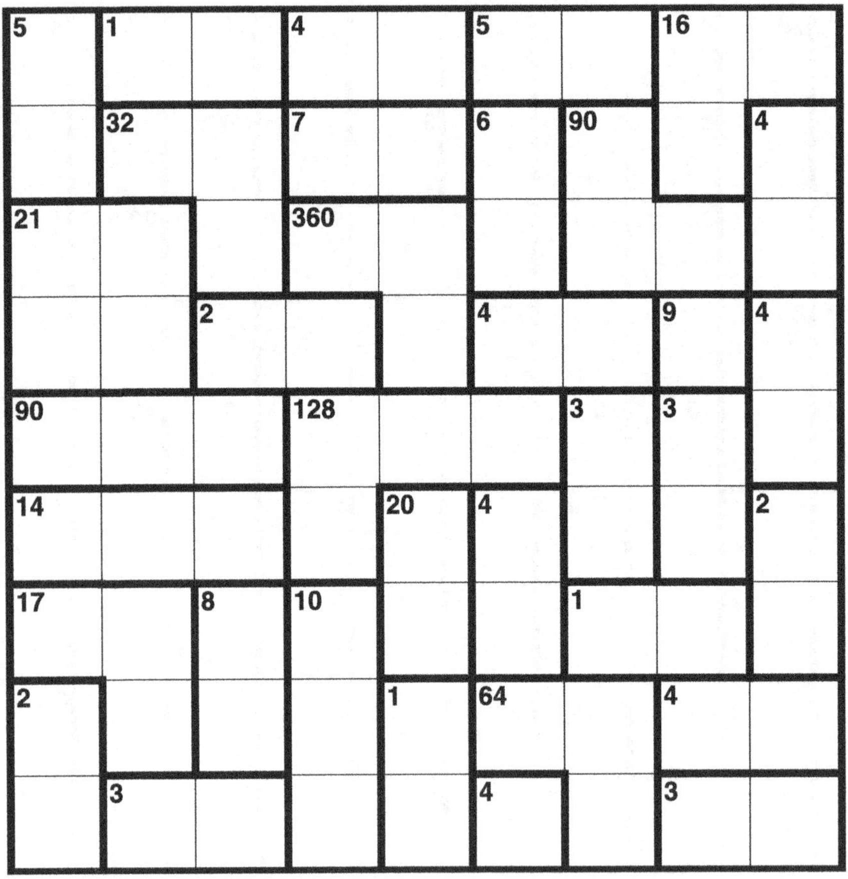

SOLVING TIME

Bonus No-Op Puzzle! This puzzle does not include operations after the target numbers... you must deduce them! Note that the operation for cages with 3 or more squares must be addition or multiplication.

104

+ − × ÷

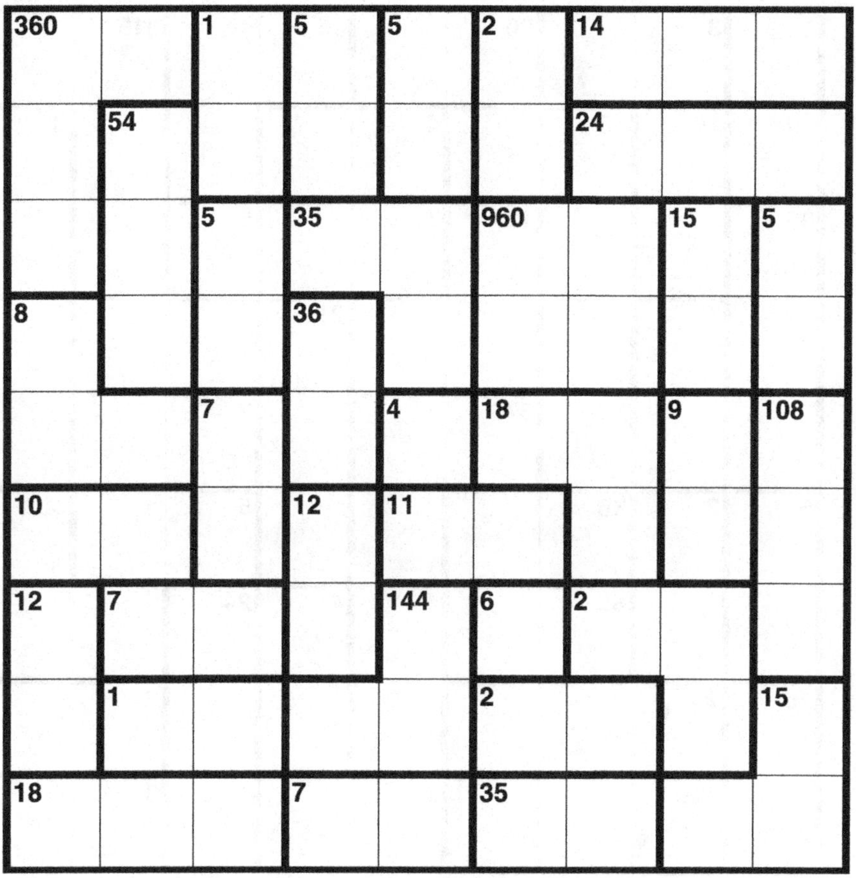

SOLVING

TIME

Bonus No-Op Puzzle! This puzzle does not include operations after the target numbers... you must deduce them! Note that the operation for cages with 3 or more squares must be addition or multiplication.

105

+ − × ÷

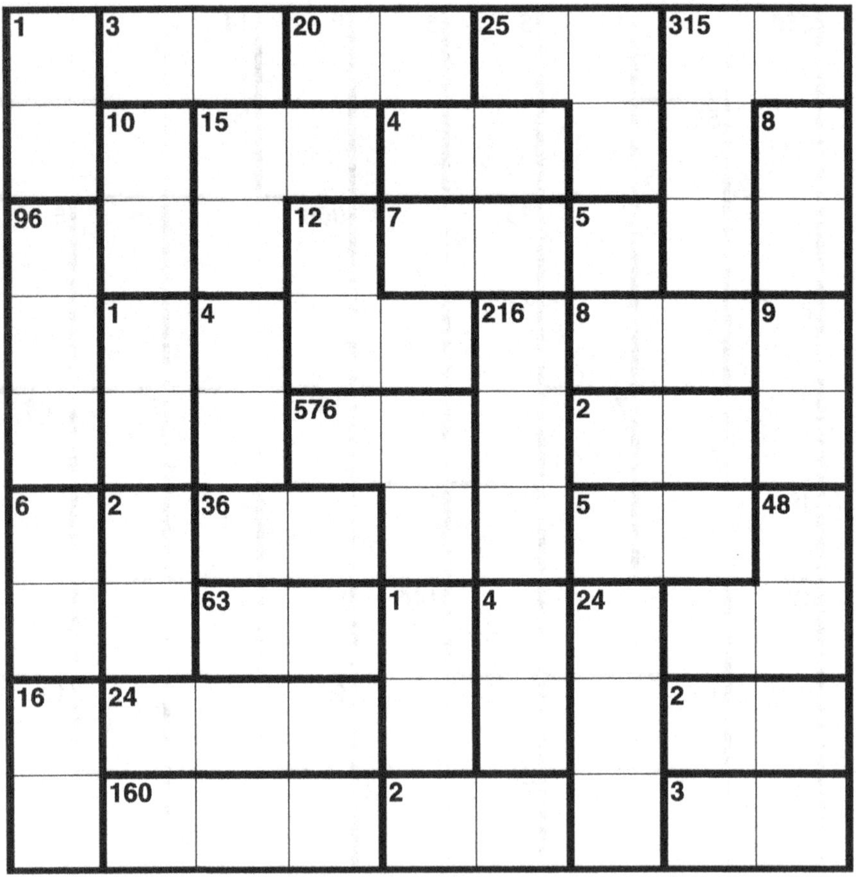

SOLVING

TIME

Bonus No-Op Puzzle! This puzzle does not include operations after the target numbers... you must deduce them! Note that the operation for cages with 3 or more squares must be addition or multiplication.

106

+ − × ÷

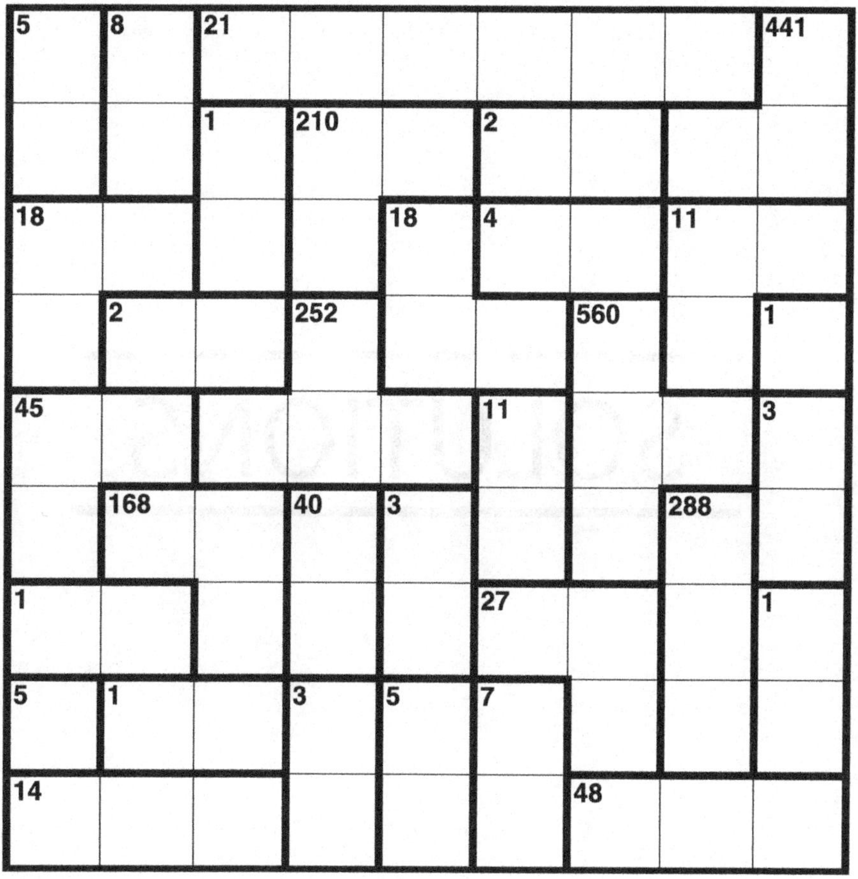

SOLVING TIME

SOLUTIONS

1

8+ 5	21+ 9	8	4	14+ 6	8+ 2	4+ 1	3	7 7
3	6+ 4	2	7	1	5	14+ 6	28+ 8	9
22+ 6	9+ 5	3	17+ 9	5+ 2	1	8	7	4
9	1	10+ 5	8	3	16+ 4	7 7	16+ 6	2
7	13+ 2	1	13+ 6	5	9	3	9+ 4	8
5+ 1	3	4	2	24+ 8	7	9	5	9+ 6
4	8	20+ 7	3 3	20+ 9	6	5	2	1
10+ 8	7	6	6+ 1	11+ 4	5+ 3	2	10+ 9	8+ 5
2	15+ 6	9	5	7	12+ 8	4	1	3

2

10+ 9	1	9+ 5	4	16+ 7	6	3	20+ 2	17+ 8
5+ 3	10+ 8	2	24+ 1	9+ 4	5	6	7	9
2	13+ 6	7	9	5+ 1	4	5	11+ 8	3
10+ 8	2	3 3	5	9	5+ 1	4	19+ 6	7
12+ 5	8+ 3	24+ 9	7	8	19+ 2	6+ 1	4	6
7	5	11+ 6	3	2	9	8	1	6+ 4
13+ 4	9	17+ 1	8	9+ 6	3	12+ 7	5	2
10+ 6	4	8	12+ 2	8+ 5	15+ 7	12+ 9	3	6+ 1
8+ 1	7	4	6	3	8	11+ 2	9	5

3

14+ 8	6	11+ 5	23+ 7	4	3	9	3+ 1	2
9+ 3	5	6	9+ 1	5+ 2	6+ 4	24+ 7	8	9
1	13+ 9	4	8	3	2	12+ 5	7	13+ 6
19+ 6	8+ 3	2	15+ 9	1	12+ 8	4	5 5	7
9	4	3	17+ 6	5	8+ 7	1	5+ 2	13+ 8
6+ 2	3+ 1	7	4	23+ 6	9	8	3	5
4	2	11+ 8	3	29+ 7	23+ 5	6	9	1 1
12+ 5	7	11+ 9	2	8	9+ 1	3	10+ 6	4
16+ 7	8	1	5	9	6	2	7+ 4	3

4

18+ 6	4	9+ 3	6+ 5	1	17+ 8	9	9+ 2	7
5+ 3	8	6	10+ 1	9	6+ 2	11+ 4	7	6+ 5
2	13+ 6	7	12+ 9	3	4	14+ 8	5	1
16+ 9	5+ 3	2	11+ 8	11+ 6	14+ 7	5	1	12+ 4
7	21+ 9	5	3	4	1	2	14+ 6	8
6+ 5	7	13+ 4	8+ 6	2	22+ 9	8+ 1	8	3 3
1	3+ 2	9	19+ 4	8	5	7	9+ 3	6
12+ 4	1	8	7	14+ 5	6	23+ 3	9	2
8	6+ 5	1	9+ 2	7	3	10+ 6	4	9

5

12+ 9	6+ 5	1	15+ 8	7	5+ 3	2	35+ 6	10+ 4
3	17+ 8	4	2	6+ 1	5	18+ 9	7	6
16+ 2	4	12+ 5	3	10+ 9	1	6	8	7 7
8	2	7	1 1	10+ 4	6	3	9	8+ 5
19+ 4	9	6	13+ 7	18+ 2	8	5+ 1	5	3
12+ 5	5+ 3	2	6	8	18+ 7	4	10+ 1	9
7	10+ 1	9	10+ 4	6	2	5 5	11+ 3	8
7+ 1	14+ 6	8	14+ 5	8+ 3	9	11+ 7	4	5+ 2
6	10+ 7	3	9	5	12+ 4	8	2	1

6

10+ 6	21+ 2	17+ 9	8	4+ 3	1	5+ 4	12+ 5	7
4	5	11+ 3	13+ 7	15+ 9	6	1	10+ 8	2
10+ 2	4	8	5	1	10+ 3	7	22+ 6	9
8	3	3+ 1	11+ 2	14+ 6	14+ 9	5	7	5+ 4
14+ 5	6	2	9	8	15+ 7	7+ 3	4	1
9	1	12+ 7	13+ 4	5	8	12+ 2	4+ 3	9+ 6
17+ 7	9	5	6 6	4	2	8	1	3
1	7 7	10+ 4	11+ 3	7+ 2	5	15+ 6	17+ 9	8
11+ 3	8	6	1	7	4 4	9	7+ 2	5

7

11+ 2	9	10+ 4	6	12+ 8	11+ 1	7	8+ 3	5
6+ 5	1	17+ 8	9	4	3	13+ 6	7	16+ 2
4+ 1	7 7	8+ 3	5	22+ 9	4	12+ 8	3+ 2	6
3	8+ 2	6	12+ 7	5	9	4	1	8
21+ 4	13+ 8	5	8+ 3	9+ 7	2	7+ 1	6	13+ 9
8	7+ 6	1	2	3	18+ 7	5	14+ 9	4
9	12+ 4	11+ 2	10+ 8	1	6	5+ 3	5	14+ 7
13+ 7	5	9	1	14+ 6	8	2	4	3
6	3	11+ 7	4	7+ 2	5	17+ 9	8	1 1

8

12+ 9	13+ 4	7+ 1	15+ 5	7	3	13+ 6	10+ 8	2
3	9	6	9+ 8	1	2	5	11+ 4	7
16+ 2	8	19+ 3	7	9+ 4	1	16+ 9	5	6 6
6	15+ 7	8	9	19+ 5	4	2	4+ 3	1
12+ 7	10+ 1	9	4 4	8	6	4+ 3	11+ 2	18+ 5
5	3 3	13+ 7	6	10+ 2	8	1	9	4
21+ 4	6	2	1	3	5	21+ 8	7	9
12+ 1	2	5	5+ 3	15+ 9	11+ 7	4	6	11+ 8
13+ 8	5	4	2	6	16+ 9	7	1 1	3

9

11+ 3	14+ 6	8	5+ 2	1	11+ 4	18+ 5	7	17+ 9
5	3	15+ 4	1	2	7	6	30+ 9	8
3+ 2	1	7	3	35+ 6	7+ 5	9	8	4
20+ 9	8	3	13+ 5	7	2	5+ 4	1	19+ 6
4 4	14+ 2	11+ 6	8	5	10+ 9	1	3	7
8	4	5	19+ 6	9	7+ 1	9+ 7	2	3
7+ 1	16+ 7	9	4	8	6	12+ 3	5	7+ 2
6	9	11+ 1	7	3	18+ 8	2	4	5
12+ 7	5	11+ 2	9	7+ 4	3	8	7+ 6	1

10

14+ 8	6	14+ 9	5	3+ 1	16+ 4	7	5+ 3	2
4+ 3	1	11+ 4	7	2	5	23+ 9	6	8
9+ 7	21+ 3	2	4	5	1	6	25+ 8	9
2	19+ 4	11+ 8	3	10+ 7	11+ 6	5	14+ 9	1
11+ 4	9	6	1 1	3	21+ 8	10+ 2	5	7
1	10+ 2	14+ 3	6	4	9	8	12+ 7	5
6	8	5	11+ 2	9	12+ 7	4 4	4+ 1	3
14+ 5	7 7	18+ 1	9	8	2	13+ 3	4	6
9	12+ 5	7	14+ 8	6	3	7+ 1	2	4

11

9+ 5	4	15+ 6	8 8	1− 2	18+ 3	9	17+ 1	7
15+ 7	2	4	5	3	6	26+ 8	9	10+ 1
6	12+ 7	16+ 3	7− 1	8	9	2	5	4
16+ 3	5	8	3− 9	6− 1	4− 4	7	4− 2	6
9	5− 1	5	6	7	8	7+ 3	4	10+ 2
4	6	6− 9	3	7+ 5	2	6− 1	7	8
11+ 8	18+ 9	7	13+ 2	6	1	4	3 3	8+ 5
1	5− 8	2	5− 4	9	7 7	11+ 5	6	3
2	3	6− 1	7	4 4	11+ 5	6	1− 8	9

12

3+ 1	2	15+ 7	5+ 3	14+ 4	9	14+ 8	6	3− 5
6 6	3	5	2	22+ 9	1	11+ 4	7	8
1− 3	2− 7	15+ 6	5	8	4 4	12+ 1	9	2
4	5	9	13+ 6	3− 2	1− 3	4− 7	9+ 8	17+ 1
1− 8	2− 6	4 4	7	5	2	3	1	9
9	8	6+ 2	4	12+ 1	5	9+ 6	3	7
3− 2	8− 1	5− 3	8	6	13+ 7	18+ 9	5	4
5	9	9+ 8	8− 1	4− 7	6	8+ 2	4	3− 3
3− 7	4	1	9	3	13+ 8	5	2	6

13

6− 8	2	20+ 5	6	3	2− 9	7	8+ 4	4− 1
16+ 9	7	23+ 8	2 2	6	5+ 1	4	3	5
10+ 3	4− 9	7	8	11+ 4	2	5	1	8+ 6
7	5	20+ 9	4	5− 8	3	8− 1	18+ 6	2
1− 2	7+ 1	3	7	1− 5	6	9	8	4
1	3	6+ 2	4− 5	5− 7	2− 4	6	17+ 9	17+ 8
2− 6	8	4	1	2	20+ 7	3	5	9
15+ 5	10+ 4	6	6− 9	10+ 1	8	6− 2	19+ 7	3
4	6	1 1	3	9	5	8	2	7

14

28+ 7	9	5	18+ 2	6	4	12+ 1	8	3
4− 8	4	7	6	7− 9	2	5 5	4+ 3	1
9+ 6	1	2− 2	3 3	5− 8	16+ 7	9	18+ 4	5
1 1	2	4	12+ 8	3	12+ 5	7	9	13+ 6
4− 9	24+ 3	6	4	3− 7	11+ 1	8	5	2
5	8	4− 3	7	4	18+ 6	2	8− 1	9
9+ 3	7	6+ 1	5	1− 2	9	2− 4	6	4− 8
2	11+ 5	1− 8	9	1	3	17+ 6	16+ 7	4
4	6	8− 9	1	5 5	8	3	2	7

15

25+ 5	6	11+ 7	4	8− 9	1	13+ 8	3	2
8	21+ 3	1	6	5	4	2	19+ 9	7
6	12+ 1	12+ 4	8	10+ 7	19+ 5	9	2 2	3
7	4	17+ 8	9	3	7− 2	5	6− 1	14+ 6
23+ 1	10+ 2	3	5	2− 6	9	20+ 4	7	8
4	9	3− 5	2	8	7	3	12+ 6	1
9	17+ 7	2 2	7+ 3	4	6	20+ 1	8	5
2	8	15+ 9	6− 7	3+ 1	11+ 3	6	5	5− 4
8+ 3	5	6	1	2	8	3− 7	4	9

16

13+ 3	6	4− 2	11+ 9	12+ 7	4	15+ 5	1	20+ 8
8− 9	4	6	2	10+ 8	1	3	5	7
1	9 9	9+ 4	2− 5	2	19+ 8	6	4− 7	3
5− 8	3	5	7	6 6	9	2	3− 4	1
13+ 6	7	4+ 3	4− 8	4− 9	5	14+ 1	6+ 2	4
7+ 2	5	1	4	8+ 3	6	7	19+ 8	9
11+ 4	8− 1	9	3 3	5	20+ 7	8	15+ 6	2
7	17+ 2	8	7+ 6	3− 1	3	16+ 4	9	1− 5
13+ 5	8	7	1	4	2	9	3	6

17

9+ 2	7	6+ 5	1	21+ 8	6	1− 9	6+ 3	11+ 4
17+ 9	5+ 2	3	2− 5	7	13+ 4	8	1	6
8	6− 3	9	7	6 6	5	4	2	1
3− 3	6	7 7	15+ 4	9	2	4− 1	14+ 8	4− 5
19+ 7	8	4	13+ 3	4+ 2	1	5	6	9
17+ 4	9	2	8	1	3− 3	6	2− 5	4− 7
6+ 1	4	15+ 6	9	9+ 5	10+ 8	2	7	3
5	10+ 1	8	11+ 6	4	10+ 7	3	7− 9	2
1− 6	5	1	2	3	2− 9	7	12+ 4	8

18

4− 8	4	11+ 7	8+ 6	6+ 1	2	15+ 9	4− 3	20+ 5
12+ 1	9	4	2	39+ 5	3	6	7	8
11+ 4	2	9+ 6	3	9	9+ 8	1	5 5	7
6	1	4+ 3	14+ 9	4	12+ 7	5	6− 8	2
5− 3	8	1	5	7	8+ 6	2	16+ 4	9
2− 7	5	3+ 2	1	6	12+ 4	17+ 8	9	3
12+ 9	3	9+ 5	4	8	1	7	8+ 2	7+ 6
18+ 2	7	9	18+ 8	3	19+ 5	4 4	6	1
5 5	2− 6	8	7	2	9	3	3− 1	4

19

2− 1	17+ 8	9	10+ 6	4	1− 3	2	12+ 7	5
3	14+ 6	8	9+ 1	17+ 7	39+ 4	3− 9	7+ 5	2
9+ 7	4+ 1	3	8	9	5	6	14+ 2	4 4
2	9 9	9+ 5	18+ 3	1	6	8	4	11+ 7
6− 8	2	4	9	6	7	5 5	3− 3	1
9+ 5	11+ 7	1	6− 2	8	9	3− 4	6	3
4	3	9+ 2	2− 7	5	8	1	15+ 9	6
15+ 6	1− 5	7	4 4	1− 2	2− 1	3	9+ 8	17+ 9
9	4	1− 6	5	3	9+ 2	7	1	8

20

1 1	15+ 5	8	2	2− 6	4	10+ 3	16+ 9	7
5− 3	15+ 9	14+ 2	7	5	6	1	14+ 8	4
8	6	5+ 4	4− 5	1	5− 7	13+ 9	2	3− 3
16+ 5	7	1	25+ 8	9	2	4	10+ 3	6
4	12+ 3	15+ 9	4+ 1	8	1− 5	6	7	10+ 2
7	2	6	3	5− 4	9	12+ 5	7+ 1	8
15+ 2	1	5	4	3	16+ 8	7	6	14+ 9
15+ 9	19+ 8	3− 3	6	7	1	6+ 2	4	5
6	4	7	7− 9	2	11+ 3	8	6+ 5	1

21

1 1	30× 3	27+ 9	6	8	28× 7	4	84× 2	4− 5
5	2	4	20+ 8	3	9	7	6	1
540× 9	1	2 2	5	22+ 4	6	40× 8	21× 3	7
112× 2	4	1	3	1− 6	8	5	13+ 7	3÷ 9
8	3− 9	6	18× 2	7	4	1	5	3
7	3− 5	8	1	9	7+ 2	5+ 3	2÷ 4	24× 6
2÷ 6	56× 7	4− 3	18+ 9	1	5	2	8	4
3	8	7	4	5	5− 1	6	8− 9	4÷ 2
10+ 4	6	70× 5	7	2	3÷ 3	9	1	8

22

1− 2	1	3− 6	3	16+ 9	1− 7	8	135× 5	1− 4
14+ 4	8	9+ 2	7	6	1	3	9	5
24+ 9	2	12+ 5	12+ 4	3− 8	1− 3	42× 6	7− 1	21× 7
6	9	7	2	5	4	1	8	3
5− 8	13+ 5	4	6	18× 2	9	7	2− 3	1
3	4	3÷ 9	4− 5	1 1	4÷ 8	2	84× 7	2− 6
2− 5	7	3	1	17+ 4	6	5− 9	2	8
6− 1	2÷ 3	7− 8	216× 9	7	10× 5	4	6	2 2
7	6	1	8	3	2	18+ 5	4	9

23

630× 7	5	36× 9	4	1	4÷ 8	2	3− 3	6
1− 4	3	1− 7	6− 9	5− 6	1	2÷ 8	7+ 5	2
5	6	8	3	63× 9	7	4	7− 2	40× 1
5− 3	8	24× 1	6	9+ 4	2	21+ 7	9	5
3− 6	9	4	5− 7	2	3	5	5− 1	8
2÷ 1	2	18× 3	1− 8	7	180× 5	9	6	28× 4
2 2	1	6	3− 5	8	9	12× 3	4	7
26+ 9	7	2	9+ 1	5	4	20+ 6	8 8	3÷ 3
8	11+ 4	5	2	3	6	1	7	9

24

3÷ 1	3	40× 4	5	192× 8	6	5+ 2	2− 7	1− 9
4− 5	1	2	13+ 7	6	4	3	9	8
2− 4	6	13+ 8	3	2	2− 7	9	4− 1	5
2− 6	17+ 8	7+ 1	4	8− 9	2− 5	21× 7	3	5− 2
8	9	6 6	2	1	3	1− 4	5	7
54× 9	2	19+ 3	1	7	8	120× 5	6	4 4
3	16+ 7	9	80× 8	5	2	1	4	18× 6
5− 2	2− 5	7	54× 9	4 4	1	2− 6	8	3
7	1− 4	5	6	3÷ 3	9	4÷ 8	2	1

25

84× 3	13+ 8	5	21× 7	3− 6	8− 1	9	2÷ 4	2
4	2 2	3	1	9	16+ 5	5− 6	1− 7	8
7	1− 5	6	9+ 4	3	8	1	14+ 2	8− 9
3÷ 2	21× 3	54× 9	5	42× 7	6	4	8	1
6	7	2	3	1	16+ 4	5− 8	14+ 9	5
576× 9	4	8	2	5	7	3	7+ 1	6
24+ 8	6	6− 1	36× 9	4	42× 2	7	270× 5	3
1	9	7	18+ 8	2	3	5 5	6	3− 4
4− 5	1	2− 4	6	8	7− 9	2	3	7

26

5− 8	3÷ 3	9	2− 7	5	2× 1	2	16+ 6	4
3	39+ 7	4÷ 8	24× 6	4− 9	5	1	4	2
15+ 6	9	2	4	9+ 7	5− 8	15× 3	5	1
9	5	20+ 7	8	2	3	3− 4	1	3− 6
12+ 7	4	2− 1	5	7+ 6	13+ 2	8	3	9
4	8	3	2 2	1	22+ 7	6	18+ 9	2− 5
1	6	1− 5	6− 3	2÷ 8	4	9	2	7
1− 2	1	4	9	2÷ 3	6	2− 5	7	8 8
3− 5	2	5− 6	1	36× 4	9	7	24× 8	3

27

4− 1	1− 5	567× 9	1− 7	8	14+ 3	48× 6	2	4
5	6	7	9	2÷ 3	2	4	8× 8	1
1− 7	8	1− 3	2	6	5	11+ 1	3− 4	18× 9
11+ 8	3	21+ 6	1− 5	4	1	9	7	2
5+ 2	8+ 7	4	3	5− 1	6	560× 8	4− 9	5
3	1	8	24× 4	7 7	5− 9	2	5	14+ 6
72× 9	2	45× 1	6	24+ 5	4	7	18× 3	8
4	9	5	1	2	5− 8	3	6	21× 7
48× 6	4	2	8	9	2− 7	5	1	3

28

4− 6	2	36× 4	9	8− 1	12+ 3	3− 5	1− 8	7
135× 3	17+ 8	1− 6	7	9	1	2	9+ 4	5
9	3	5− 2	11+ 4	7	8	12+ 1	5	6
5	6	7	1− 2	1− 3	4	8− 9	1	96× 8
63× 7	9	24× 1	3	2÷ 4	3− 5	8	6	2
20× 4	5	3	1	8	13+ 6	7	72× 2	9
1	2− 7	5	8	11+ 2	9	54× 6	3	4
4− 8	4	23+ 9	6	1− 5	5− 2	3	7 7	2− 1
1− 2	1	8	5 5	6	7	36× 4	9	3

29

12× 4	1728× 6	22+ 5	2	8− 9	1	16+ 7	8	162× 3
3	8	96× 2	5	7	1− 4	1	6	9
9	4	6	8	3	5	2÷ 2	1	2− 7
16+ 6	2 2	15× 3	13+ 4	1	1− 9	8	19+ 7	5
2	5	1	168× 6	8	420× 7	9	3	2÷ 4
8	2− 7	9	1	4− 6	3	4	5	2
6+ 5	2− 1	4	7	2	11+ 8	3	15+ 9	6 6
1	3	63× 7	9	720× 4	6	5	2	7− 8
16+ 7	9	11+ 8	3	10× 5	2	6	4	1

30

3− 7	8− 9	1	96× 8	4	3	3− 2	5	20+ 6
4	6+ 1	18+ 2	8+ 5	3	2520× 7	9	6	8
18× 3	2	4	210× 7	6	8	5	7− 1	9 9
1	3	9	2÷ 2	5	1− 6	7	8	1120× 4
6	180× 4	3	1	18+ 9	14+ 2	8	7	5
9	5	192× 8	4	7	1	6	2	3
3− 5	1− 8	7	6	2	3÷ 9	3	4÷ 4	1
2	336× 6	270× 5	9	8 8	9+ 4	8+ 1	3	18+ 7
8	7	6	3÷ 3	1	5	4	9	2

31

2÷ 1	2	2÷ 4	2÷ 3	6	480× 8	5	112× 7	10206× 9
6+ 5	1	2	2− 9	7	4	3	8	6
192× 4	6	32× 8	4− 5	9	3÷ 3	1	2	7
24+ 6	8	1	4	21+ 5	7	14× 2	9	3
2	252× 4	5 5	17+ 8	3	9	7	7+ 6	1
3	9	7	6	64× 8	2	4	4− 1	5
8	5	3− 9	11+ 7	2	4− 1	3− 6	3	16+ 4
4− 7	3	6	2	24× 1	5	720× 9	4	8
2− 9	7	2− 3	1	4	6	8	5	2

32

162× 9	14+ 2	7	3	18+ 4	8	6	14+ 5	1
3	6	2	15+ 5	9	1	4÷ 4	18+ 7	8
210× 7	5	6	72× 8	5− 3	10+ 4	1	9	2
4− 2	3÷ 3	2÷ 4	9	8	6	4− 5	1	2− 7
6	1	8	5− 2	7	2− 3	5− 9	4	5
13+ 4	63× 7	9	5− 1	6	5	1− 8	54× 2	3 3
8	5− 4	11+ 5	6	12+ 1	2 2	7	3	9
1	9	11+ 3	4	5	9+ 7	2	2− 8	6
3− 5	8	1	7	2	3÷ 9	3	2− 6	4

33

324× 9	5 5	2÷ 4	2	7− 1	8	30× 6	12+ 7	3
3	6	63× 7	25+ 8	9	1	5	2	1− 4
2− 7	2	9	1	8	432× 6	3	4	5
5	63× 9	1	7+ 4	3	56× 7	8	6	2 2
6− 2	7	2− 3	5	11+ 4	8− 9	1	48× 8	6
8	16+ 3	6	7	5	2	144× 4	1	17+ 9
2− 6	13+ 8	2	3	22+ 7	4	9	8+ 5	1
4	8× 1	8	9	6	3− 5	2	3	7
3− 1	4	30× 5	6	42× 2	3	7	1− 9	8

34

2÷ 2	4− 9	5	8× 8	1	1− 7	6	672× 4	3
4	5+ 2	3	8− 1	30× 5	6	7	8	7− 9
2− 7	5	3− 6	9	2÷ 4	6− 8	4− 1	2÷ 3	2
108× 3	4	9	10+ 7	8	2	5	6	4− 1
9	2− 6	4	3	4− 7	4− 1	6− 8	2	5
8 8	23+ 7	2÷ 2	4	3	5	16+ 9	1	6
9+ 5	8	1	21+ 2	54× 6	18+ 3	4 4	21+ 9	7
1	3	7	6	9	4	6× 2	5	8 8
5− 6	1	8	5	2	9	3	3− 7	4

35

4− 4	8	14+ 6	162× 3	9	1− 5	5− 2	189× 1	7
60× 5	2	8	10+ 1	6	4	7	9	3
3÷ 9	6	1− 3	7	168× 8	4− 1	5	2− 2	4
3	315× 9	4	2	7	5− 6	1	15+ 5	8
7	5	6× 1	6	3	1− 8	9	4 4	2
6+ 2	3	39+ 5	9	4	7	8	6	8− 1
1	64× 4	2	8	15× 5	3	72× 6	168× 7	9
378× 6	1	7	1− 4	2÷ 2	18× 9	3	8	30× 5
1− 8	7	9	5	1	2	4	3	6

36

5− 6	1	20× 5	4	3÷ 3	9	24+ 8	16+ 7	2
22+ 3	6	28+ 9	8	4	3÷ 1	5	108× 2	7
5	8	7	1− 2	1	3	4	6	9
23+ 9	4	3	3− 5	8	2− 2	7	7+ 1	6
7	5− 2	5− 6	1	360× 9	4	108× 3	120× 5	8
1− 1	7	384× 2	6	5	8	9	4	3
2	3÷ 3	4	22+ 9	7	6	1 1	5− 8	20× 5
4− 4	9	8	7 7	3÷ 6	3− 5	2	3	1
8	9+ 5	1	3	2	378× 7	6	9	4

37

6− 9	3− 4	280× 7	24× 3	144× 8	60× 6	5	2	5− 1
3	7	1	8	2	9	1− 4	5	6
2÷ 4	2	8	10+ 6	21× 1	3	7	3− 9	12+ 5
7− 8	1	5	4	9 9	5+ 2	3	6	7
39+ 5	9	4	7	6	8	10+ 1	1− 3	2
7+ 1	6	7− 2	2− 5	7	168× 4	9	5− 8	3
3÷ 2	3− 5	9	9+ 1	3	7	6	3− 4	288× 8
6	8	3− 3	18× 9	5	2÷ 1	2	7	4
4− 7	3	6	2	9+ 4	5	9+ 8	1	9

38

3÷ 9	2− 7	144× 6	1	3	8	40× 5	2− 4	2
3	5	7− 9	10+ 7	7− 8	1	4	2	192× 6
504× 7	9	2	3	11+ 5	6	12+ 1	8	4
8	20+ 6	5	9	3÷ 2	1− 4	7	3	1
480× 4	6+ 1	3	2	6	5	2− 9	7	11+ 8
6	4	14× 7	160× 5	15+ 9	2	8× 8	1	3
5	2	1	8	4	4− 7	3	1− 6	16+ 9
11+ 2	5− 3	8	4	8+ 1	18+ 9	6	5	7
1	8	10+ 4	6	7	3	16+ 2	9	5

39

135× 5	9	16+ 2	8	21× 3	7	2− 4	5− 6	1
3	3− 7	4	6	63× 9	1	2	50× 5	8 8
3− 9	2÷ 4	14+ 6	1	7	5− 8	3	2	5
6	8	7	18+ 2	5− 1	15× 5	5− 9	4	72× 3
4÷ 2	10+ 5	9	7	6	3	7− 1	8	4
8	3	10+ 5	4	15+ 2	63× 9	7	1	6
12+ 7	2	1	5	8	2− 4	6	3÷ 3	9
4	1	81× 3	9	1− 5	6	3− 8	25+ 7	2
1 1	2− 6	8	3	2− 4	2	5	9	7

40

1− 7	8	168× 3	1− 4	5	60× 6	10+ 9	1	10+ 2
12+ 4	6	8	27× 9	3	2	5	7	1
5− 3	2	7	1	288× 8	9	4	30× 6	5
8	21+ 4	5	2	1	3	6	20+ 9	7
14× 2	7	5− 1	6	4− 9	5	3÷ 3	4	2÷ 8
3− 9	8+ 3	336× 6	7	2÷ 2	7− 8	1	7+ 5	4
6	5	7− 9	8	4	1	1− 7	2	3 3
8− 1	9	2	5 5	3− 7	4	8	2÷ 3	6
4− 5	1	1− 4	3	13+ 6	7	19+ 2	8	9

41

2÷ 2	1	30× 6	5	15+ 3	4	8	16+ 7	9
2− 6	8	9 9	56× 7	8+ 5	3	60× 2	15+ 1	4
5− 4	11+ 9	2	8	1	5	6	3	7
9	6− 7	1	8+ 3	4	3− 8	5	4− 6	2
21× 3	2÷ 2	4	1	1− 6	7	360× 9	5	8
1	2− 5	12× 3	4	1− 9	3÷ 2	252× 7	15+ 8	6
7	3	80× 5	2	8	6	4	9	1
17+ 5	4	8	1− 6	7	54× 9	6+ 1	2	3
8	22+ 6	7	9	2	1	3	1− 4	5

42

14+ 5	4	180× 9	2− 1	5+ 3	2	21+ 6	7	8
2÷ 6	5	4	3	29+ 8	1− 1	2	2− 9	7
3	14× 1	5	7	9	2÷ 8	4	14+ 6	2
1	7	7+ 3	4− 4	5	22+ 9	80× 8	2	6
2	3	1	8	7	6	5	18+ 4	9
20+ 4	6− 2	8	108× 6	11+ 1	7	10+ 9	5	3 3
7	14+ 6	2	9	2÷ 4	3	1	16+ 8	5
9	8	30× 6	5	2	140× 4	7	3	4÷ 1
24+ 8	9	7	3÷ 2	6	5	3÷ 3	1	4

43

14+ 9	324× 6	3	4 4	10+ 2	1	7	3− 5	8
5	9	2	4− 3	7	64× 8	4	3× 1	6 6
24+ 8	7	19+ 9	6	4	2	19+ 5	3	1
7	4 4	17+ 5	4− 9	8− 1	10+ 3	6	8	2− 2
2	3	8	5	9	7	9+ 1	6	4
2÷ 4	8	1	35× 7	5	2÷ 6	3	2	3÷ 9
2÷ 1	2	39+ 4	8	6	5	9	7	3
5− 6	1	84× 7	2	12+ 3	17+ 9	8	9+ 4	5
2− 3	5	6	1	8	2÷ 4	2	16+ 9	7

44

6− 1	54× 6	9	1− 3	12+ 2	5	56× 7	8	4 4
7	1− 3	4	2	5	13+ 6	1− 8	9	5− 1
40× 5	8	16+ 3	4	9	7	2÷ 2	1	6
2÷ 4	2	1− 7	17+ 9	2÷ 8	2÷ 1	6+ 5	18+ 6	3
3÷ 3	35× 7	6	8	4	2	1	2− 5	9
9	5	14× 2	7	11+ 1	4	6	3	80× 8
4− 8	4	14+ 5	1	54× 6	9	10+ 3	7	2
4− 2	1	8	1− 6	21× 7	216× 3	9	17+ 4	5
6	10+ 9	1	5	3	8	4	2	7

45

30× 1	10+ 4	6	10+ 7	3	14× 2	29+ 9	5	8
6	5	4÷ 8	2	360× 9	1	7	1− 3	4
378× 3	2	9	2− 1	8	7	15+ 6	4	5
15+ 4	7	1− 2	3	5	14+ 8	1	3− 6	9
2	6 6	3	36× 4	1	9	5	28+ 8	7
9	20× 1	4	1− 5	14× 2	96× 3	8	7	6
17+ 8	9	5	6	7	4	1− 3	2	6× 1
2− 5	24× 8	2− 7	9	10+ 4	6	18× 2	1	3
7	3	1	2− 8	6	9+ 5	4	9	2

46

3÷ 2	5− 6	1	1− 4	5	24× 3	8	16+ 7	9
6	8 8	15+ 3	2	9	1	20× 5	4	35× 7
17+ 8	2− 9	20× 4	105× 7	12+ 6	6× 2	3	1	5
9	7	5	3	2	2− 8	11+ 1	6	4
9+ 3	1	18+ 9	5	4	6	5− 7	2	48× 8
5	3	6	2− 9	7	5− 4	17+ 2	8	1
2− 7	5	16+ 2	8	21× 1	9	4	3	6
4÷ 1	4	1− 8	6	3	7	4− 9	5	1− 2
2− 4	2	7	14+ 1	8	5	3− 6	9	3

47

84× 2	21+ 9	7	5	4− 4	3÷ 1	15+ 8	3− 3	6
6	7	5− 9	4	8	3	5	1	13+ 2
12× 4	3÷ 2	6	18+ 7	5	4− 9	1	8	3
3	1	1− 4	6	7 7	5	144× 2	9	8
4− 9	5	3	1− 8	2− 6	4	14× 7	2	1
4− 5	4÷ 4	1	9	7− 2	8 8	1− 3	13+ 6	16+ 7
1	144× 6	8	3	9	15+ 2	4	7	5
15+ 7	8	8+ 2	1	6× 3	6	1080× 9	5	4
5− 8	3	5	2	1	7	6	4	9 9

48

1344× 3	8	6 6	13+ 7	4	2	13+ 1	4− 9	5
7	4− 9	8− 1	1− 5	6	8	4	18+ 3	2÷ 2
8	5	9	2− 3	1	6	7	2	4
8− 1	5+ 2	3	3− 8	5	24+ 7	9	168× 4	48× 6
9	1− 4	2÷ 2	1	126× 7	3	5	6	8
60× 5	3	2÷ 8	4	2	5− 1	6	7	24+ 9
2	6− 1	120× 5	6	9	4 4	5− 3	8	7
6	7	4	14+ 2	3	9	8	6+ 5	1
1512× 4	6	7	9	13+ 8	5	6+ 2	1	3

49

35× 1	5	17+ 2	8	7	27+ 4	9	6	3
7	1	2÷ 8	3÷ 9	3	5	2÷ 2	4	16+ 6
2÷ 8	13+ 6	4	2− 3	5	270× 9	1	7	2
4	7	2÷ 3	16× 1	8	6	5	18× 2	9
540× 9	360× 8	6	1− 7	1	2	21+ 4	15× 3	5
3	9	5	6	2	8	7	1	4 4
5	4	21+ 9	2	6	24× 1	3	8	4032× 7
10+ 2	3÷ 3	1	4	1260× 9	21× 7	1− 6	5	8
6	2	7	5	4	3	8	9	1

50

30× 6	26+ 5	9	1− 2	3	21+ 1	12× 4	28+ 7	8
5	8	4	10+ 1	9	2	3	17+ 6	7
504× 8	9	20× 5	4	3− 7	3	18× 2	1	6
6− 1	7	2÷ 3	6	4	5	9	8	14+ 2
7	3÷ 1	22+ 8	9	5	4	1− 6	2	3
1− 2	3	42× 1	40× 5	8	6	7	17+ 4	9
3	2÷ 2	6	7	5− 1	504× 8	40× 5	9	4
19+ 9	4	1− 2	3	6	7	8	15× 5	6+ 1
4	6	7 7	4÷ 8	2	9	1	3	5

51

2016× 8	6	7	84× 3	10+ 5	2	10+ 9	2÷ 4	10+ 1
6	108× 9	15+ 5	7	4	3	1	8	2
1	4	2	8	7− 9	2− 5	3	22+ 6	7
35× 7	3	24+ 4	6	2	24+ 9	8	1	1− 5
5	1	8	24× 2	3	7	6	9	4
72× 9	7 7	6	4	9+ 8	1	12+ 5	2	2÷ 3
4	40× 5	8− 9	1	6− 7	2− 8	2	3	6
2	8	9+ 3	4− 5	1	6	140× 4	7	17+ 9
3	2	1	9	17+ 6	4	7	5	8

52

8× 1	8	3÷ 2	8+ 5	3	22+ 6	9	7	12+ 4
2− 7	9	6	3÷ 3	1	12+ 4	15+ 2	4÷ 8	5
3− 8	5	63× 7	9	4	1	6	2	3
48× 6	4 4	27× 9	1	10× 2	3	7	3− 5	8
4	3÷ 2	3	28× 7	5	13+ 8	24× 1	16+ 9	6
2	6	1344× 8	4	2− 7	5	3	1	10+ 9
4− 5	7	4	6	9	2 2	8	72× 3	1
9	12+ 3	1	8	18+ 6	7	5	4	5− 2
2− 3	1	7+ 5	2	21+ 8	9	4	6	7

53

6+ 1	5	2÷ 4	96× 6	8	2	135× 9	3	5– 7
3– 6	9	8	5– 3	3– 7	4	5	135× 1	2
252× 9	4	7	8	12× 1	6	2	5	3
2– 3	1	5 5	18+ 4	1– 6	112× 7	8	2	9
4÷ 2	5– 6	1	9	5	4– 3	7	1– 8	120× 4
8	3÷ 3	9	5	13+ 2	4÷ 1	4	7	6
11+ 4	4– 2	6	504× 7	3	8	8– 1	9	5
7	11+ 8	3	2	9	11+ 5	6	2– 4	7– 1
2– 5	7	2÷ 2	1	4	3÷ 9	3	6	8

54

108× 9	1	2	168× 7	11+ 4	2– 6	3– 8	5	2– 3
6	7– 2	8	3	7	4	8– 9	1	5
35× 7	9	2÷ 4	2	432× 6	8	120× 5	3	1
1	12+ 3	5	4	5+ 2	9	42× 7	6	8
5	2– 4	6	23+ 9	3	11+ 1	2	8	3– 7
24× 3	1– 8	4– 7	6	7+ 5	2	18+ 1	9 9	4
4	7	3	8	8× 1	2– 5	6	2	9
2	19+ 5	11+ 9	1	8	3	1– 4	1– 7	6
8	6	1	315× 5	9	7	3	2÷ 4	2

55

147× 7	3	7– 9	2	18+ 6	8	1	160× 4	5
3– 5	7	2– 4	6	1– 2	1	3	8	19+ 9
8	10× 5	2	3÷ 3	168× 7	6	4	9	1
72× 2	6 6	35× 5	9	21× 3	7	13+ 8	1	4
4	9	7	4– 5	1	360× 2	18+ 6	3	16+ 8
19+ 6	14+ 2	1	1– 7	8	4	9	5	3
3	8	7+ 6	1	180× 4	9	5	9+ 7	2
1	4	15+ 3	8	9	105× 5	4– 2	6	84× 7
9	7– 1	8	4	5	3	7	2	6

56

42× 6	2– 4	10× 2	5	1	1– 8	7	6– 9	3
7	6	216× 4	9	22+ 5	1	14+ 3	8	1– 2
15× 5	18+ 8	6	3÷ 2	9	7	5– 4	3	1
3	2	8	6	2– 7	5	9	4÷ 1	4
15+ 1	5	3	4	2	3÷ 9	48× 6	56× 7	8
18+ 8	9	8+ 7	1	2÷ 6	3	2	4	18+ 5
9+ 2	1	14+ 9	18+ 8	3	18+ 4	5	6	7
4	3	5	7	8 8	12+ 6	1	18× 2	9
63× 9	7	1	3	4	2	8	11+ 5	6

57

18+ 3	3− 7	4	54× 1	9	6	15+ 2	40× 8	5
7	6− 9	42× 6	2− 4	12× 1	3	8	5	6× 2
8	3	7	6	4	90× 5	9	2	1
3÷ 6	14+ 5	11+ 9	2	70× 7	224× 8	3− 4	1	3
2	8	3÷ 3	9	5	4	4− 1	15+ 6	13+ 7
4 4	1	7− 8	21× 3	2	7	5	9	6
3− 5	2	1	7	24× 3	8− 9	3− 6	4− 4	8
3− 9	6	3− 2	13+ 5	8	1	3	4− 7	5− 4
3− 1	4	5	8	4− 6	2	7 7	3	9

58

49× 1	7	16+ 6	2− 8	3− 2	5	1− 9	1− 4	3
7	9	1	6	15+ 5	3	8	2− 2	4
2− 8	6	2÷ 2	1	7	5− 9	4	14+ 3	5
2÷ 2	4	5 5	5− 3	8	17+ 1	7	9	6
1− 4	5	18+ 9	2	42× 6	7	5− 3	8	1− 1
3÷ 9	3	12× 4	7	7− 1	8	1− 5	6	2
4− 6	2	3	45× 5	9	11+ 4	1	2− 7	1− 8
15× 3	1	23+ 8	5− 9	4	4− 2	6	5	7
5	8	7	12× 4	3	6	1− 2	1	9 9

59

2− 7	12× 1	6	2	22+ 5	9	8	15+ 3	4
5	17+ 3	9	1 1	2÷ 2	4	1− 6	7	8
2÷ 1	2	5	18+ 9	4	7− 8	12× 3	1− 6	7
4− 3	7	6× 2	5	72× 9	1	4	2− 8	6
24× 4	6	3	48× 8	1	2− 5	7	13+ 9	5+ 2
15+ 9	5	1	6	8	5− 7	2	4	3
2− 6	5− 4	2÷ 8	63× 7	3	2÷ 2	1	45× 5	9
8	9	4	3	42× 7	2÷ 6	3− 5	2	6+ 1
4÷ 2	8	3− 7	4	6	3	8− 9	1	5

60

5− 1	5− 2	3− 6	224× 8	7	1− 9	12+ 4	3	5
6	7	3	4	3− 2	8	6+ 5	8− 1	9
72× 8	9	1− 2	3	5	8+ 7	1	2− 6	4
14+ 9	5	196× 4	7	25+ 6	1	1− 3	560× 8	2
24× 3	8	7	6	9	4	2	5	5− 1
2÷ 2	4	4− 9	2× 1	2− 3	5	8 8	7	6
4− 7	3	5	2	1	324× 6	9	2÷ 4	8
1− 5	5− 1	72× 8	9	9+ 4	3	6	18+ 2	10+ 7
4	6	1	13+ 5	8	2	7	9	3

61

24× 8	3	2÷ 2	4	2− 5	7	3− 9	6	4÷ 1
18+ 6	1	30× 5	18× 9	1− 3	2	15+ 7	8	4
3	9	6	2	1− 8	1− 5	4	17+ 1	7
5− 4	13+ 5	8	1	7	3− 6	3	5− 2	9
9	3− 6	3	1− 5	32× 1	4	8	7	2 2
14+ 7	5− 4	9	6	12+ 2	3÷ 3	1	5 5	2− 8
2	6− 7	1	1− 8	4	45× 9	5	15+ 3	6
5	4− 8	4	7	6	2÷ 1	2	9	3
10+ 1	2	7	3÷ 3	9	48× 8	6	20× 4	5

62

3− 8	13+ 6	384× 2	36× 1	4	9	15+ 3	5	7
5	7	6	4	2÷ 2	1	16× 8	135× 3	9
252× 7	9	13+ 3	8	2− 6	4	1	2	5
4	3	7	4− 5	1	8 8	378× 6	8− 9	2÷ 2
48× 3	10× 2	5	48× 6	8	12+ 7	9	1	4
2	8	1	21+ 3	9	5	7	11+ 4	15+ 6
15+ 6	4− 5	21+ 8	9	2− 3	2÷ 2	4	7	1
9	1	4	5− 7	5	6× 3	2	17+ 6	8
4÷ 1	4	9	2	210× 7	6	5	8	3

63

4÷ 1	4	1− 8	9	252× 2	7	6	3	11+ 5
24+ 6	9	13+ 7	5− 3	8	4− 1	5	12+ 2	4
9	3− 5	6	4÷ 1	4	72× 8	3	7	2
1− 5	8	6− 1	7	3− 6	9	4÷ 2	3− 4	10+ 3
4	240× 2	5	6	9	2÷ 3	8	1	7
24× 3	14+ 1	2	4	105× 5	6	1− 7	8	16+ 9
8	7	4	40× 5	3	17+ 2	9	6	1
1− 2	3	3÷ 9	8	7	9+ 4	4− 1	5	6
1− 7	6	3	2÷ 2	1	5	288× 4	9	8

64

2÷ 1	2	15× 5	3	144× 4	6	34+ 7	9	8
3− 7	4	1− 3	2	6	13+ 5	8	60× 1	9
3÷ 9	3	140× 7	4	5	288× 8	6	2	1
5− 3	8	6 6	1	9	4	5	5− 7	2
3− 2	5	7+ 1	6	1− 8	7	21+ 9	2− 4	15× 3
2÷ 4	1− 1	2	56× 8	6− 7	9	3	6	5
8	288× 9	4	7	1	3÷ 3	2÷ 2	11+ 5	6
1− 5	378× 6	8	4− 9	12× 2	1	4	4− 3	7
6	7	9	5	3	2	1 1	2÷ 8	4

65

72× 9	4	2	11+ 1	4− 7	3	1680× 5	8	6
1− 6	15+ 8	288× 4	5	3	2	7	7− 9	4÷ 1
5	7	8	9	22+ 6	13+ 1	3	2	4
5− 3	108× 6	1− 1	2	8	9	13+ 4	20× 5	378× 7
8	3	6	7	1	5 5	9	4	2
1260× 7	5	9	4	4÷ 2	8	20+ 6	8+ 1	3
8+ 2	1	2− 5	12× 3	4	6	8	7	9
4	7− 2	3	21+ 8	4− 9	3− 7	6× 1	6	3− 5
1	9	7	6	5	4	1− 2	3	8

66

2÷ 3	6	112× 4	2	7	2− 5	3− 1	1− 8	9
23+ 6	8	2	8− 1	9	7	4	13+ 3	5
504× 8	9	4− 1	6 6	13+ 4	3	2− 5	7	2
9	1− 4	5	4÷ 8	2	6	16+ 7	1	3
7	5	3 3	1− 4	4− 6	2	8	10+ 9	1
2− 4	5− 2	7	3	8− 1	9	2160× 6	5	8
2	21+ 7	8	45× 5	5− 3	4÷ 1	9	288× 4	6
10+ 5	1	6	9	8	4	1− 3	2	3− 7
1	3	2− 9	7	13+ 5	8	2	6	4

67

28× 4	1008× 8	7	6	10+ 9	1	50× 5	11+ 3	2
7	1	3	9 9	56× 4	5	2	56× 8	6
2− 8	6	2÷ 4	2	7	3÷ 3	9	1	15+ 5
19+ 6	4	8	90× 5	2	9	15× 1	7	3
42× 2	9	4÷ 1	4	14+ 8	6	3	5	7
3	7	15+ 6	3− 8	5	2÷ 2	4	8− 9	1
3− 5	2	9	3÷ 3	1	23+ 4	7	6	21+ 8
8+ 1	5	2	6− 7	144× 3	8	6	4	9
17+ 9	3	5	1	6	1− 7	8	2÷ 2	4

68

336× 4	2	6	8− 9	5 5	6− 1	7	3− 8	10+ 3
7	24× 6	4	1	216× 8	3	9	5	2
2÷ 1	2− 3	6− 9	2÷ 2	4	21+ 7	8	6	5
2	1	3	1− 4	432× 6	8	315× 5	9	7
3− 5	3− 7	13+ 2	3	12+ 1	9	3− 6	2÷ 4	8
8	4	5	6	9	2	3	8+ 7	1
27+ 9	3− 5	8	22+ 7	6× 3	30× 6	4÷ 1	2÷ 2	4
3	9	7	8	2	5	4	5− 1	6
6	7− 8	1	2− 5	7	2÷ 4	2	3÷ 3	9

69

3− 8	84× 4	3	7	7+ 2	5	26+ 9	5− 1	6
5	1− 3	2	3− 1	4	6	7	5− 8	8− 9
84× 6	2	17+ 9	22+ 8	5	7 7	4	3	1
336× 2	7	8	6	3	4÷ 1	6+ 5	13+ 9	4
3	8	60× 6	5	10+ 9	4	1	5− 7	2
7	20× 5	4	2	1	19+ 9	2÷ 3	6	1− 8
5− 4	9	1	5− 3	8	2	3÷ 6	20× 5	7
10+ 1	6 6	13+ 5	20+ 9	7	8	2	4	8+ 3
9	1	7	4	2÷ 6	3	4÷ 8	2	5

70

3÷ 2	56× 8	8× 4	1	2÷ 3	6	21+ 7	17+ 5	9
6	7	2	120× 5	9	1	4	3	15+ 8
23+ 4	9	3	8	10+ 5	12+ 7	6− 2	6	1
9 9	4	6	14× 7	1	5	8	1− 2	3
1− 7	6	1	2	4	5− 8	2− 3	2− 9	1− 5
6+ 1	4÷ 2	8	3− 9	6	3	5	7	4
5	8+ 3	2− 9	36× 6	6− 8	2	10+ 1	4÷ 4	1− 7
11+ 8	5	7	3	2	216× 4	9	1	6
3	4− 1	5	28× 4	7	9	6	4÷ 8	2

71

60× 6	6× 1	26+ 5	8	4	3÷ 3	18× 9	13+ 2	7
5	6	1− 8	7	9	1	2	16+ 3	4
2	4− 7	3	168× 4	6	384× 8	1	9	4− 5
30× 3	5	2	1	7	6	8	4	9
3− 1	5− 8	4− 9	5	3÷ 3	2− 2	4	1680× 7	6
4	3	7 7	216× 6	1	14+ 9	5	8	6− 2
21+ 9	4	6	3	14+ 2	5	11+ 7	1	8
8	13+ 9	4	2	5	7	3	30× 6	3÷ 1
5− 7	2	18+ 1	9	8	2− 4	6	5	3

72

105× 3	5	7	9+ 4	2	15+ 6	8	1	504× 9
48× 6	1	18× 9	2	3	20+ 4	5	7	8
40× 5	8	3÷ 3	9	2− 6	7	4	2÷ 2	1
2	4	2− 5	7	8	8+ 3	10+ 1	3− 9	6
3− 7	3− 6	192× 8	16+ 5	4	1	9	1− 3	2
4	9	2	8	45× 1	5	126× 3	6	7
7− 1	2	6	3	4− 5	9	20+ 7	8	1− 4
8	189× 7	4× 4	1	9	2 2	96× 6	5	3
9	3	1	1− 6	7	8	2	1− 4	5

73

2− 9	7	24× 8	3	2÷ 2	4	3÷ 6	13+ 1	5
2− 5	3− 6	5− 4	9	9+ 8	1	2	11+ 3	7
7	9	6+ 1	72× 6	12+ 5	3	4	8	15+ 2
1− 2	1	5	4	3	240× 8	17+ 9	7	6
7+ 1	2÷ 2	504× 9	7	6	5	8	18+ 4	3 3
6	4	18+ 7	8	1− 1	2	6× 3	5	9
14+ 8	3	6	5	3− 7	3− 9	1	2	4− 4
3	240× 5	2	2÷ 1	4	6	315× 7	9	8
4 4	8	3	2	63× 9	7	5	6× 6	1

74

1− 5	4− 3	7	108× 2	6	392× 4	216× 9	1 1	13+ 8
6	45× 5	9	1	2	7	8	3	4
5− 2	1− 4	5	9	7	48× 8	3	39+ 6	1
7	3− 9	20× 4	5	5− 1	6	2	8	12+ 3
1− 9	6	24× 8	3	3− 5	2	42× 1	4	7
8	8+ 7	1	16+ 4	9	3	6	5	2
2÷ 4	2	48× 6	8	3÷ 3	1	7	9	45× 5
5+ 3	1	3÷ 2	6	19+ 8	18+ 5	4	7	9
1	5− 8	3	7	4	9	3− 5	2	6 6

75

35× 1	120× 8	3	5	21+ 9	2	7	2− 6	4
7	5	15+ 4	2	6	3	17+ 8	9	10+ 1
2 2	18+ 7	5− 1	3	14+ 8	20+ 6	5	4	9
8	3	6	6− 1	2	4	9 9	5	4− 7
4÷ 4	1	23+ 9	7	3− 5	8	6× 6	7+ 2	3
20× 5	4	8	6	12+ 7	4− 9	1	3	2
162× 9	2 2	2− 7	4	1	5	17+ 3	8	6
3	3− 6	5	13+ 9	4	11+ 7	56× 2	40× 1	8
6	9	6− 2	8	3	1	4	7	5

76

3÷ 1	3	10× 2	588× 7	360× 5	8	13+ 9	4	22+ 6
168× 6	2− 8	5	4	9	15× 1	7+ 2	3	7
4	6	21+ 8	3	7	5	1	2	9
7	8− 1	4	9	12+ 8	2− 6	3	15+ 5	2
6− 2	9	8+ 1	5	3	4	42× 6	7	8
8	2− 4	6	2	1	16+ 3	2− 7	9	2− 5
10+ 5	2	1− 7	6	4	9	160× 8	5− 1	3
3	4− 5	9	48× 8	11+ 2	7	4	6	1 1
19+ 9	7	3	1	6	2	5	2÷ 8	4

77

270× 9	3− 5	1− 8	7	1− 6	15+ 3	1	2÷ 2	4
5	8	8− 9	1	7	11+ 4	6	3	2
6	12+ 3	4	5	72× 9	7	1− 2	7− 1	8
2÷ 2	4 4	126× 7	6	8	11+ 5	3	18+ 9	1
4	15+ 2	6	3	5	1	3− 7	8	3÷ 9
6− 1	7	14+ 5	9	4÷ 2	8	4	16+ 6	3
7	3÷ 9	3	14+ 8	2÷ 1	2	45× 5	4	6
9+ 8	1	2	4	2÷ 3	6	9	5 5	35× 7
18× 3	6	1	72× 2	4	9	1− 8	7	5

78

9072× 9	3	7+ 5	2	21+ 1	8	7	24× 6	4
6	8	7	168× 3	7+ 5	2	1	4	24+ 9
2− 5	28× 4	8	7	10+ 9	1	20+ 2	3	6
3	7	270× 9	5	6	24+ 4	8	2	1
4÷ 1	9+ 6	4÷ 2	8	7 7	3	4	9	5
4	2	1	2− 6	16+ 8	9	135× 5	35× 7	3
16× 8	1	2÷ 3	4	2	6	9	5	1− 7
2	180× 5	6	10+ 9	3− 4	7	3	1	8
7 7	9	4	1	14+ 3	5	6	4÷ 8	2

79

4− 1	5	1− 2	3	4− 4	8	84× 7	6	3÷ 9
13+ 6	7	189× 9	3− 8	5	27× 1	4÷ 4	2	3
4÷ 8	2	7	42× 6	3	9	1	35× 5	2− 4
14+ 5	9	3	7	3− 6	2÷ 4	8	1	2
12+ 2	1	14+ 4	3− 5	9	2÷ 6	3	7	1− 8
9	12+ 3	1	2	1− 8	90× 5	24× 6	4	7
3	6	5	4	7	2	9	7− 8	1
3− 7	2÷ 4	14+ 8	1− 1	2	2− 3	5	15+ 9	6
4	8	6	17+ 9	1	7	2 2	2− 3	5

80

1− 3	4	1− 2	7− 9	384× 6	8	39+ 5	8+ 7	1
18+ 7	1	3	2	8	1− 5	9	2− 6	1− 4
8	2	4− 1	5	7− 9	6	7	4	3
1 1	7+ 3	4	13+ 6	2	4− 7	8	17+ 5	15+ 9
504× 2	21+ 5	8	7	10+ 1	3	4	9	6
4	8	10+ 9	1	7	2	6	3	3− 5
9	7	6 6	120× 8	5	1− 4	3	11+ 1	2
1− 5	6	35× 7	3	1− 4	8− 9	1	2	8
3− 6	9	5	4 4	3	2÷ 1	2	1− 8	7

81

6– 1	2– 9	7	2– 3	2÷ 4	8	360× 2	6	5
7	5– 4	9	5	13+ 1	4– 2	6	5– 8	3
105× 5	7	3	56× 8	9	6	4÷ 4	1	2 2
2– 4	6	4– 1	7	3	2– 9	40× 5	4÷ 2	8
2÷ 3	24× 1	5	2÷ 4	2	7	8	3– 9	6
6	3	8	2÷ 2	2– 7	18+ 5	9	4	8– 1
11+ 2	14+ 8	6	1	5	4÷ 4	4– 7	3 3	9
9	15+ 5	3÷ 2	6	17+ 8	1	3	3– 7	4
8	2	13+ 4	9	6	3	35× 1	5	7

82

2– 7	3÷ 3	1	192× 6	8	4	16+ 2	9	5
5	54× 9	3	2	42× 7	4÷ 1	4	2– 6	8
160× 8	5	63× 9	7	6	2– 3	1	5– 2	2– 4
4	120× 8	5	3	1	756× 2	9	7	6
9+ 3	2	4÷ 8	17+ 5	17+ 9	6	7	4÷ 4	1
10+ 1	4	2	8	5	2– 7	18+ 6	3	9
9	13+ 6	7	4	3	5	40× 8	1	12+ 2
1– 6	7	3– 4	1	2÷ 2	1– 9	5	8 8	3
2÷ 2	1	54× 6	9	4	8	2– 3	5	7

83

1– 7	48× 6	8	15+ 1	4	2	3	5	14+ 9
6	108× 9	3	4	4– 5	6– 1	7	7– 8	2
3– 4	7	280× 2	6– 8	9	168× 6	4– 5	1	3
5– 3	8	5	2	7	4	9	6× 6	1
6+ 5	1	7	13+ 6	2	108× 9	4	126× 3	8 8
1– 9	2– 2	4	5	144× 8	3	1	7	6
8	4	45× 1	1134× 3	6	1– 7	16+ 2	9	5
6× 1	5	9	7	3	8	3÷ 6	2	28× 4
2	3	6	9	4– 1	5	2÷ 8	4	7

84

12+ 8	3	200× 5	4× 2	1	5– 4	9	1– 6	7
2÷ 6	1	8	5	2	8+ 7	432× 4	9	3
3	54× 9	6	3– 7	80× 5	1	16× 8	2	4
13+ 7	6	3÷ 9	4	8	2	1	5– 3	4– 5
2÷ 4	2	3	3– 6	9	2– 5	7	8	1
17+ 5	8	3÷ 1	3	2– 4	3÷ 6	2	2– 7	9
2÷ 1	4	1– 7	8	6	3÷ 9	3	4– 5	4– 2
2	3– 7	4	3÷ 9	3	3– 8	5	1	6
14+ 9	5	2÷ 2	1	16+ 7	3	6	4– 4	8

85

90× 3	6	5	108× 2	9	1 1	28+ 4	7	19+ 8
11+ 7	7− 1	1− 4	2÷ 3	6	70× 5	8	9	2
4	8	3	6	7	2	4− 5	1	9
11+ 8	13+ 4	1	24+ 9	3− 3	6	3− 2	5	2− 7
1	2	6	7	5− 4	9	15+ 3	8	5
3− 6	9	2	8	20+ 5	7	1 1	4	3÷ 3
90× 9	700× 5	7	4	8	2÷ 3	6	1− 2	1
2	1− 7	8	5	2× 1	13+ 4	9	3	16+ 6
5	3÷ 3	9	1	2	1− 8	7	6	4

86

2− 5	10+ 9	2÷ 4	72× 8	6− 7	1	3÷ 6	30× 2	3
7	1	8	9	2÷ 6	3	2	5	4− 4
11+ 3	5 5	13+ 9	4	3÷ 2	6	15+ 1	7	8
8	1− 2	36× 1	6	5− 9	4	7	15× 3	5
5− 1	3	6	280× 7	8	11+ 5	13+ 9	4	2 2
6	4− 7	3	5	1	2	4	7− 8	2− 9
20+ 9	6	5	6× 2	1− 4	15+ 8	3 3	1	7
2÷ 4	4÷ 8	2	1	3	7	40× 5	3− 9	5− 6
2	3− 4	7	3	4− 5	9	8	6	1

87

1− 3	2	72× 1	9	8	7+ 4	18+ 5	6	7
23+ 7	2− 3	5	6− 8	2	1	3− 6	9	9+ 4
9	7	4 4	2	3− 6	15+ 8	3× 3	1	5
11+ 4	1− 5	6	84× 3	9	7	1	384× 8	2
1	6	252× 9	7	4	15+ 5	2	3	8
4÷ 8	4	7	5− 6	1	2	2− 9	15× 5	3
2	64× 1	8	13+ 5	3	6	7	3− 4	16+ 9
54× 6	9	2	4	5	3÷ 3	2÷ 8	7	1
3− 5	8	21× 3	1	7	9	4	2 2	6

88

12× 3	4	14+ 2	5	1	1− 6	7	72× 8	9
224× 7	8	4	6	16+ 5	2	3÷ 3	9	1
4÷ 2	21× 7	9+ 5	2− 1	3	9	23+ 6	720× 4	4− 8
8	3	1	5− 2	7	1− 5	9	6	4
10+ 1	9	3	2− 7	2÷ 2	4	8	5	6
3÷ 6	2	2− 8	9	4	3÷ 1	3− 5	105× 3	7
3− 4	1	6	1− 8	9	3	2	7 7	5
1− 5	6	23+ 9	4	1− 8	7	4÷ 1	6+ 2	3
4− 9	5	7	3	2− 6	8	4	1	2 2

89

22+ 6	9	10+ 2	5	2÷ 4	12× 1	3	1− 7	8
7	14+ 1	5	3	8	4	3− 6	9	3− 2
7+ 3	168× 7	8	16+ 9	20+ 6	2	4	6+ 1	5
4	6	1	7	420× 5	8	2	3	3÷ 9
144× 9	4	6	2	7	2− 5	18+ 1	8	3
8	2	54× 9	6	1	3	5	4	7 7
18+ 5	8	1− 3	8× 1	378× 9	6	7	2÷ 2	4
2÷ 2	5	4	8	21× 3	7	9 9	5− 6	1
1	10+ 3	7	2÷ 4	2	2160× 9	8	5	6

90

16+ 5	8	7− 2	56× 7	5+ 3	3− 4	16+ 1	9	6
2− 6	3	9	4	2	1	13+ 7	5	1− 8
4	3÷ 9	3	2	6 6	3− 5	8	1	7
3÷ 9	2− 7	5	3÷ 3	48× 8	6	120× 4	2÷ 2	1
3	192× 6	8	1	2− 7	9	5	11+ 4	1− 2
4÷ 2	5 5	4	72× 9	1	8	6	7	3
8	1− 2	6+ 1	5	16+ 4	7	2÷ 3	6	36× 9
8+ 7	1	14+ 6	8	5	108× 2	9	3	4
1	17+ 4	7	6	3÷ 9	3	2	3− 8	5

91

26880× 8	5	6	54× 9	2	3	12+ 1	7	4
25+ 5	8	2÷ 2	1	12× 3	4	9072× 6	9	7
1	2	7	30+ 5	9	6	4	24× 3	8
6	3	9	4	5 5	98× 2	7	216× 8	1
2	1	4	6	40× 8	7	240× 3	5	9
30+ 9	6	8	7	1	5	2	4	3
21× 7	3− 4	2− 1	3	14+ 6	8	26+ 9	2	11+ 5
3	7	10× 5	2	3− 4	9	8	9+ 1	6
5− 4	9	5− 3	8	7	6+ 1	5	6	2

92

252× 7	6	16× 2	8	1	24× 3	18+ 5	4	9
6	27× 9	1	12+ 5	7	8	84× 4	3	2÷ 2
4− 5	1	3	1− 2	4− 9	3− 6	56× 8	7	4
1	2÷ 2	4	3	5	9	7	15+ 8	1− 6
1− 2	3	1008× 9	7	8	36× 4	1	6	5
21+ 8	4	19+ 7	6	2	1	9	9+ 5	3
4	5	6	19+ 9	3	7	36× 2	1	9+ 8
39+ 9	7	8	4	6	5	3	126× 2	1
16+ 3	8	5	7+ 1	4	2	6	9	7

93

4− 3	1− 8	9	5− 6	1	11+ 2	4	2− 5	7
7	18× 9	2	1	27+ 8	4	5	2÷ 6	3
2÷ 8	35× 5	7 7	2− 4	9	6	3÷ 2	2− 3	1
4	1	2÷ 3	2	2− 7	9	6	320× 8	5
10+ 1	7	6	4− 9	20× 4	5	5− 3	882× 2	8
9	4÷ 4	1	5	13+ 2	3	8	7	2− 6
3÷ 6	2	2÷ 8	3	5	17+ 1	7	9	4
3− 5	90× 6	4	5− 8	3	7	9	12+ 1	2
2	3	5	1− 7	6	8 8	4÷ 1	4	9

94

4− 6	2	90× 5	3÷ 3	1	72× 9	8	14+ 4	7
11+ 7	4÷ 1	2	9	9+ 4	5	6 6	40× 8	3
3	4	5− 7	2	23+ 9	6	5	1	8 8
1	1− 3	4	2− 5	7	8	3− 9	6	4− 2
4− 4	8	3÷ 9	42× 1	6× 2	3	2− 7	5	6
360× 8	9	3	7	11+ 6	4	1	7+ 2	5
16+ 2	5	5− 1	6	40× 8	5− 7	9+ 3	6− 9	3− 4
9	21+ 7	6	2÷ 8	5	2	4	3	1
5	6	8	4	2− 3	1	2	2− 7	9

95

15+ 3	5	1	2	4	21+ 8	6	63× 9	7
5− 6	1152× 4	9 9	2÷ 1	2	7	45× 5	3	560× 8
1	9	4	8	3÷ 6	2	3	7	5
3− 9	6	2− 7	5	7− 8	1	140× 4	2	108× 3
15+ 7	8	16+ 2	6	18+ 9	3	1	120× 5	4
17+ 4	2	8	126× 3	1	5	7	6	9
8	3	6	7	5	19+ 9	19+ 2	4	11+ 1
8+ 5	1	84× 3	4	7	6	9	8	2
2	2− 7	5	3÷ 9	3	4	8	5− 1	6

96

2− 5	3	96× 8	5− 2	7	18+ 9	4	12+ 1	3− 6
5− 7	2	4	3	192× 1	5	6− 8	6	9
15+ 9	245× 7	5	6 6	8	3	2	4	1
6	20× 4	7	16× 1	15+ 2	8	17+ 9	5	3
4÷ 1	5	2	8	4	6	3	1008× 9	7
4	10+ 6	1	4− 9	5	15+ 2	7	4− 3	8
144× 8	9	3	2− 4	6	1	5	7	2
2	144× 1	315× 9	5	3÷ 3	168× 7	6	2÷ 8	4
3	8	6	7	9	4	10× 1	2	5

97

504× 8	9	5− 2	7	48× 6	20× 5	4	8+ 1	3
4− 6	7	14+ 1	5	8	3÷ 3	9	4	10+ 2
2	2÷ 1	12× 4	8	378× 9	6	7	3	5
10+ 1	2	3	28× 4	42× 7	22+ 8	5	9	24× 6
9	3− 5	7	1	3	2	8 8	3÷ 6	4
9+ 5	8	13+ 6	3	4	6− 7	1	2	16+ 9
4	19+ 6	5	18+ 9	2	1	5− 3	8	7
4− 3	1− 4	8	6	6+ 5	5− 9	4− 2	2− 7	7− 1
7	3	11+ 9	2	1	4	6	5	8

98

2− 4	2	24× 1	1890× 7	9	6	5+ 3	3− 8	5
13+ 6	7	3	8	5	15+ 1	2	2÷ 4	6− 9
11+ 7	4	10+ 9	1	6	8	1− 5	2	3
4− 5	1260× 9	7	1− 3	6− 8	2	4	7+ 1	6
1	5	4	2	7 7	8+ 3	72× 9	1− 6	4÷ 8
3 3	20+ 6	5	9	1	4	8	7	2
7− 9	5− 3	8	11+ 4	2	1− 7	6	9+ 5	1
2	96× 8	6	5	18+ 4	9	1 1	3	1764× 7
7− 8	1	2	2÷ 6	3	5	7	9	4

99

16+ 9	4− 3	7	280× 5	8	6+ 4	1	3÷ 6	2
5	2	5− 3	8	7	1	216× 4	9	6
22+ 7	6× 1	6	2÷ 3	10+ 9	80× 8	5	2	1− 4
4	9	2	6	1	63× 7	5− 8	3	5
2÷ 3	1− 5	4	2÷ 2	3÷ 6	9	63× 7	17+ 8	1
6	336× 7	4− 5	1	2	10+ 3	9	11+ 4	8
8	6	1	252× 9	4	5	2	7	3÷ 3
2÷ 1	20+ 4	8	7	1− 3	2	1− 6	5	9
2	8	36× 9	4	30× 5	6	11+ 3	1	7

100

540× 6	9	147× 7	12+ 2	5	4	1	224× 8	162× 3
5	2	3	48× 8	6	1	4	7	9
7− 9	16+ 5	1	7	4 4	112× 8	2	72× 3	6
2	3	270× 5	9	1728× 8	7	6	4	35× 1
15+ 1	8	6	4	9	18+ 2	3	5	7
4	7	32× 2	6	27× 1	3	5	1− 9	8
3	1	4	25+ 5	7	9	8	13+ 6	2
1− 8	4	9	3	3÷ 2	6	8+ 7	1	5
7	14+ 6	8	1	17+ 3	5	9	2− 2	4

101

[35] 5	7	[576] 8	[30] 3	2	[7] 1	4	[19] 9	6
[6] 1	8	9	5	[162] 3	2	[42] 6	7	4
3	[14] 1	[2] 4	8	9	6	[5] 7	2	[210] 5
2	9	[4] 5	1	[60] 4	3	[72] 8	6	7
[21] 7	4	[10] 3	6	1	5	9	[32] 8	[6] 2
9	5	[14] 2	7	[1] 6	[1] 8	1	4	3
[3] 6	2	[3] 7	4	5	9	[7] 3	1	[1] 8
[2] 4	[2] 6	[11] 1	2	8	[2] 7	5	3	9
8	3	[3] 6	9	[28] 7	4	[10] 2	5	1

102

[90] 6	[4] 2	8	[105] 3	5	[8] 1	[12] 7	4	[4] 9
3	[8] 4	2	7	[6] 6	9	[3072] 8	1	5
5	[12] 7	1	8	4	2	6	[22] 9	[1] 3
[8] 9	1	4	[126] 6	7	3	5	8	2
1	[15] 3	[15] 6	9	[126] 2	7	[56] 4	[13] 5	8
[4] 8	5	[1701] 3	[17] 1	9	[4] 4	2	7	[7] 6
2	9	7	4	[11] 8	[3] 5	[3] 3	[4] 6	1
[2] 4	6	9	5	3	8	1	2	[3] 7
[15] 7	8	5	2	[5] 1	6	[6] 9	3	4

103

[5] 9	[1] 4	5	[4] 7	3	[5] 6	1	[16] 8	2
4	[32] 1	8	[7] 9	2	[6] 7	[90] 5	6	[4] 3
[21] 2	8	4	[360] 5	9	1	6	3	7
5	6	[2] 2	4	8	[4] 3	7	[9] 9	[4] 1
[90] 3	5	6	[128] 2	1	8	[3] 9	[3] 7	4
[14] 1	2	7	8	[20] 5	[4] 9	3	4	[2] 6
[17] 7	3	[8] 9	[10] 6	4	5	[1] 2	1	8
[2] 8	7	1	3	[1] 6	[64] 2	4	[4] 5	9
6	[3] 9	3	1	7	[4] 4	8	[3] 2	5

104

[360] 9	5	[1] 6	[5] 8	[5] 1	[2] 2	[14] 3	7	4
4	[54] 2	5	3	6	1	[24] 9	8	7
2	9	[5] 7	[35] 1	5	[960] 8	4	[15] 6	[5] 3
[8] 1	3	2	[36] 4	7	5	6	9	8
6	1	[7] 8	9	[4] 4	[18] 3	7	[9] 5	[108] 2
[10] 3	7	1	[12] 5	[11] 2	9	8	4	6
[12] 5	[7] 4	3	7	[144] 8	[6] 6	[2] 1	2	9
7	[1] 8	9	6	3	[2] 4	2	1	[15] 5
[18] 8	6	4	[7] 2	9	[35] 7	5	3	1

105

3	2	6	4	5	8	9	1	7
4	9	2	6	3	7	8	5	1
6	1	7	2	4	3	5	9	8
8	6	5	3	2	9	1	7	4
2	7	1	8	9	4	6	3	5
1	5	4	9	8	6	7	2	3
5	3	9	7	6	1	4	8	2
9	8	3	1	7	5	2	4	6
7	4	8	5	1	2	3	6	9

106

8	9	3	2	1	6	4	5	7
3	1	8	6	5	4	2	7	9
4	8	9	7	3	5	1	6	2
6	4	2	9	8	7	5	3	1
9	5	1	4	7	3	8	2	6
1	6	4	5	2	8	7	9	3
2	3	7	8	6	1	9	4	5
5	7	6	1	9	2	3	8	4
7	2	5	3	4	9	6	1	8

Made in the USA
Monee, IL
13 September 2026

62976605R00079